CHEMICAL PROCESSING
With a BASIC Computer

by
Eric Reichter

BASIC SCIENCE PRESS

Palos Verdes Estates
California

"Everything that changes must be divisible,
for every change is from something to something.

When a thing is at the goal of its change
it is no longer changing.

When both itself and all of its parts are at the
starting point of its change it is not changing.

Part of that which is changing must then be at the starting point
and part at the goal.

As a whole, it cannot be in both or in neither."

—Aristotle

FOREWORD

Chemical Processing is the Philosopher's Stone which transmutes gross substances into precious ones. It is the literal foundation of civilization, furnishing materials of all descriptions.

Chemical Processing is a modern technology; a new science. Its laws and methods have evolved in barely more than a century, though its origins date to prehistoric times. The ancients used its techniques for the refining of metals and the production of glass. Yet, it remained an unreliable art, its essence unexplained. The science of today might be said to begin in 1828 with the synthesis of an "organic" material, urea, by Friedrich Wohler, His discovery revolutionized thinking of what was possible and what was not.

Near the end of the Nineteenth Century, Adolf von Bayer was able to make indigo dye. Its manufacture led to the collapse of indigo plantations in India and Louisiana.

During World War I, the Central Powers were able to produce ammonia for gunpowder which made them independent of the mines of Chile. During World War II, the Axis made synthetic rubber from oil, and oil from coal.

Following World War II, technology was developed for producing chlorine and flourine compounds which do not exist in nature. Our markets have been transformed by such plastic products as Nylon, Teflon, and PVC (Poly-Vinyl-Chloride).

Chemical Processing deals with the structure of matter and its transformation. However, we do not need to look to nuclear physics for understanding. Rather, we can build upon simple measurements of heat, a knowledge of diagramatic organization, and the application of general scientific laws.

INTRODUCTION

Structure and Energy.

These are the building blocks of Chemical Processing.

By Structure, we are ordinarily merely concerned with the number and types of atoms constituting a molecule and with its freedom of movement as exhibited by being a solid, a liquid, or a gas.

It is with Energy that we must apply the weight of our analysis. Nineteenth Century theoreticians Ludwig von Helmholtz and Josiah Gibbs pioneered this effort. They found that the total energy of a substance could be divided into a "Free Energy" component, available for work and chemical processes, and an "Unavailable Energy" component, which Rudolph Clausius called Entropy.

Chemical Processing explores the forms and types of relations for molecular structure. It defines the laws which govern reactions between molecules and determines those changes which are possible and those which are not. To those who understand its capabilities and limitations, it can guide in developing new methods and products.

BASIC CONCEPTS

ARRHENIUS THEORY OF ELECTROLYTIC DISSOCIATION —
Chemical reactions require the dissociation of molecules into charged particles and the recombination of those particles to form new molecules.
— Svante August Arrhenius (1859-1927)

LAW OF ARRHENIUS — Reaction rates double or even triple with an increase of temperature of 10 degrees Centigrade.

LAW OF MASS ACTION - For any reversible reaction, when equilibrium obtains, regardless of the direction from which that equilibrium is approached, the product of the concentrations of the molecules on one side of the equation divided by the product of the concentrations on the other is equal to a constant, characteristic of the reaction, at a given temperature.
— Guldberg and Waage, 1867

PRINCIPLE OF LeCHATELIER — When a factor determining the equilibruim of a system is altered, the system tends to change in such a way as to oppose and partially annul the alteration in this factor.
— Henri Louis Le Chatelier (1850-1936)

LAW OF van't HOFF — When the temperature of a system in equilibrium is raised, the equilibrium point is displaced in the direction which absorbs heat. When the temperature is lowered, the displacement is in the direction which gives off heat.
— Jacobus Hendricus van't Hoff (1852-1911)

HESS'S LAW — The amount of heat absorbed or liberated in a chemical reaction is constant regardless of the number of stages occuring if the same original substances and end products are involved.
— Germain Henri Hess (1802-1850)

CONSERVATION OF ENERGY — The total energy of an isolated system remains constant. (Sometimes called The First Law of Thermodynamics.)
— Julius Mayer, 1842

FREE ENERGY — The potential of the system for performing external work is limited by the total energy of the system and the unavailable energy as measured by temperature and entropy.
— Josiah Gibbs (1839-1903) and
Ludwig von Helmholtz (1821-1894)

CONTENTS

THE STRUCTURE OF MATTER

ATOMIC STRUCTURE

There are 92 atoms of principal interest, as differentiated by chemical properties. These range from Hydrogen, the most abundant and lightest, to Uranium, the heaviest. These atoms have been arranged by Mendeleeff, and subsequently others, into a Periodic Chart which indicates similarity of properties.

There are additional types of atoms, present in minor amounts, as differentiated by molecular weights. These are known as Isotopes and were discovered in 1919 by F. W. Aston, an English physicist.

The Hydrogen isotope of Mass 1 has a weight of 1.0078 (relative to Oxygen with a weight of 16.000). The Hydrogen isotope of Mass 2 has a weight of 2.0141. There is a third isotope of Mass 3 which is radioactive and unstable. The relative abundance of these types is indicated by the following chart.

HYDROGEN ISOTOPES

	MASS 1	MASS 2	MASS 3
Name:	Protium	Deuterium	Tritium
Abundance:	99.985 %	0.015 %	7×10^{-14}

An additional variation in the structure of matter is the condition of molecular spin. Hydrogen is composed of para and ortho molecules distinguished by rotation in opposite directions.

VALENCE

The combining ratio of an atom with other atoms is expressed by the valence. Some atoms have a fixed valence, for which this ratio is a constant. For example, hydrogen has a valence of one so that one atom of hydrogen always combines with other atoms to form one molecule of the simplest order. About four-fifths of the atoms have a variable valence, with several possible combining ratios. Valence is thought to be related to the number of electrons in the outer shell of the atom.

The concept of valence was given by Sir Edward Frankland (1825-1899), English chemist, who wrote in 1852: "the combining power of the attracting element is satisfied by the same number of atoms of the elements which it attracts."

TABLE I. ATOMIC WEIGHTS AND NUMBERS

1	Hydrogen	H	1.0078		48	Cadmium	Cd	112.41
2	Helium	He	4.003		49	Indium	In	114.76
3	Lithium	Li	6.940		50	Tin	Sn	118.70
4	Beryllium	Be	9.02		51	Antimony	Sb	121.76
5	Boron	B	10.82		52	Tellurium	Te	127.61
6	Carbon	C	12.010		53	Iodine	I	126.92
7	Nitrogen	N	14.008		54	Xenon	Xe	131.3
8	Oxygen	O	16.000		55	Cesium	Cs	132.91
9	Fluorine	F	19.00		56	Barium	Ba	137.36
10	Neon	Ne	20.183		57	Lanthanum	La	138.92
11	Sodium	Na	22.997		58	Cerium	Ce	140.92
12	Magnesium	Mg	24.32		59	Praseodymium	Pr	140.92
13	Aluminum	Al	26.97		60	Neodymium	Nd	144.27
14	Silicon	Si	28.06		61	Illinium	Il	146.
15	Phosphorus	P	30.98		62	Samarium	Sm	150.43
16	Sulfur	S	32.06		63	Europium	Eu	152.0
17	Chlorine	Cl	35.457		64	Gadolinium	Gd	156.9
18	Argon	A	39.944		65	Terbium	Tb	159.2
19	Potassium	K	39.096		66	Dysprosium	Dy	162.46
20	Calcium	Ca	40.08		67	Holmium	Ho	163.5
21	Scandium	Sc	45.10		68	Erbium	Er	167.26
22	Titanium	Ti	47.90		69	Thulium	Tm	169.4
23	Vanadium	V	50.95		70	Ytterbuyn	Yb	173.04
24	Chromium	Cr	52.01		71	Lutecium	Lu	174.99
25	Manganese	Mn	54.93		72	Hafnium	Hf	178.6
26	Iron	Fe	55.85		73	Tantalum	Ta	180.88
27	Cobalt	Co	58.94		74	Tungsten	W	183.92
28	Nickel	Ni	58.69		75	Rhenium	Re	186.31
29	Copper	Cu	63.57		76	Osmium	Os	190.2
30	Zinc	Zn	65.38		77	Iridium	Ir	193.1
31	Gallium	Ga	69.72		78	Platinum	Pt	195.23
32	Germanium	Ge	72.60		79	Gold	Au	197.2
33	Arsenic	As	74.91		80	Mercury	Hg	200.61
34	Selenium	Se	78.96		81	Thallium	Tl	204.39
35	Bromine	Br	79.916		82	Lead	Pb	207.21
36	Krypton	Kr	83.7		83	Bismuth	Bi	209.00
37	Rubidium	Rb	85.48		84	Polonium	Po	210.
38	Stronthium	Sr	87.63		85	Astatine	At	210.
39	Yttrium	Y	88.92		86	Radon	Rn	222.
40	Zirconium	Zr	91.22		87	Francium	Fr	224.
41	Niobium	Nb	92.91		88	Radium	Ra	226.05
42	Molybdenum	Mo	95.95		89	Actinium	Ac	227.
43	Technetium	Tc	97.		90	Thorium	Th	232.12
44	Ruthenium	Re	101.7		91	Protoactinium	Pa	231.
45	Rhodium	Rh	102.91		92	Uranium	U	238.07
46	Palladium	Pd	106.7		93	Neptunium	Np	237.
47	Silver	Ag	107.88		94	Plutonium	Pu	242.

2

TABLE II.   VALENCE

Aluminum	3	Neodymium	3
Americium	3,4,5,6	Neon	0
Antimony	3,5	Neptunium	4,5,6
Argon	0	Nickel	2,3
Arsenic	3,5	Niobium	3,5
Astatine	1,3,5,7	Nitrogen	3,5
Barium	2	Osmium	2,3,4,8
Berkelium	3,4	Oxygen	2
Beryllium	2	Palladium	2,4,6
Bismuth	3,5	Phosphorus	3,5
Boron	3	Platinum	2,4
Bromine	1,3,5,7	Plutonium	3,4,5,6
Cadmium	2	Potassium	1
Calcium	2	Praseodymium	3
Carbon	2,4	Promethium (#61)	3
Cerium	3,4	Radium	2
Cesium	1	Rhodium	3
Chlorine	1,3,5,7	Rubidium	1
Chromium	2,3,6	Ruthenium	3,4,6,8
Cobalt	2,3	Samarium	2,3
Copper	1,2	Scandium	3
Curium	3	Selenium	2,4,6
Dysprosium	3	Silicon	4
Erbium	3	Silver	1
Europium	2,3	Sodium	1
Fluorine	1	Strontium	2
Francium	1	Sulfur	2,4,6
Gadolinium	3	Tantalum	5
Gallium	2,3	Technetium	6,7
Germanium	4	Tellurium	2,4,6
Gold	1,3	Terbium	3
Hafnium	4	Thallium	1,3
Helium	0	Thorium	4
Holmium	3	Thulium	3
Hydrogen	1	Tin	2,4
Indium	3	Titanium	3,4
Iodine	1,3,5,7	Tungsten	6
Iridium	3,4	Uranium	4,6
Iron	2,3	Vanadium	3,5
Krypton	0	Xenon	0
Lanthanum	3	Ytterbium	2,3
Lead	2,4	Yttrium	3
Lithium	1	Zinc	2
Lutetium	3	Zirconium	4
Magnesium	2		
Manganese	2,3,4,6,7		
Mercury	1,2		
Molybdenum	3,4,6		

MOLECULAR FORMS

It is theorized that at absolute zero temperature all substances form a perfect crystalline shape of minimum energy level
in regard to the movement of molecules.

At ordinary temperatures all substances can be classified
into those which form crystals and those which, like glass, have
no particular molecular arrangement. Glass, and other amorphous
(without form) substances have no defined melting or solidificat-
ion point.

Crystals can be found in six specific shapes:

> Cubic with three axes of equal length, all at right
> angles to each other.

> Tetragonal system with three axes, two of equal length,
> all intersecting at right angles.

> Orthorhombic system with three axes of unequal length,
> all intersecting each other at right angles.

> Monoclinic system with three axes of unequal length, two
> of which intersect obliquely, while the third
> intersects the other at right angles.

> Triclinic system with three axes of unequal length, none
> of which intersect at right angles.

> Hexagonal system with three axes of unequal length in
> one plane intersecting each other at 60 degree
> angles, and a fourth axis at right angles to the
> plane of the other three.

CHEMICAL BONDS

Molecules are held together with chemical bonds. The ionic
or electrovalent bond case has electrons completely transferred
from one atom to another. The covalent bond case has electrons
shared between atoms. The sharing of electrons constitutes a
single bond.

The hydrogen molecule, H_2, is an example of a covalent bond.
Having two of the same atoms it is thought that neither would
pull an electron from the other.

Ionic substances react easily upon contact. This may be done
by melting the substance or by dissolving it in water or other
solvent. The ions contact and the reaction can occur rapidly.

Covalent substances require a source of energy to break the
bonds. Energy can be supplied by light or heat. The Carbon-
Hydrogen (C-H) bond is stronger than the Carbon-Carbon (C-C) bond.
The C-H bond requires 87.3 Kilocalories to be broken; the C-C
bond only 58.6 Kcal.

von BAEYER'S STRAIN THEORY

Adolph von Baeyer was a pupil of Kekule' of Darmstadt and discoverer of the formula for indigo dye. His theory states that the four valencies of the carbon atom act normally in the direction of the four lines drawn drom the center to the four corners of a regular tetrahedron (thus making included angles of 109° $28'$). If the directions of the valencies are different from this normal condition, a state of intramolecular strain is set up, its magnitude depending on the angle through which the directions of the valencies are diverted.

Such modifications of structure can account for variations in the apparent bond strength.

IONIZATION

The necessity for ionization to precede chemical reactions was promoted by Svante Arrhenius in a paper published in 1887, working with the conduction of electricity in solutions.

His experiments were made with salt, sodium chloride, dissolved in water. He theorized that the salt ionized into sodium and chloride ions under the influence of the water and not of the electric current.

Arrhenius noted that the ions possessed different properties than the elements, since the chloride ions did not have the green-yellow color of chlorine gas.

The degree of ionization can be expressed by the dissociation constants or ionization constants.

CHANGE OF PHASE

Change of phase, as from a liquid to a gas, or from a solid to a liquid, requires that the material undergo a rearrangement of the molecular structure. The energy absorbed or released by such a change of phase must be considered in defining the total energy of the substance.

CLASSIFICATION OF COMPOUNDS BY COMPOSITION

In the early days of chemistry it was thought that compounds produced by life processes possesed a special life force. These were called Organic Compounds with the others known as Inorganic Compounds. In 1824, one of these Organic Compounds, Urea, was synthesized by the German Chemist, Friedrich Wohler. His discovery revolutionized concepts of chemistry.

These terms are still used today, but in a different context. The characteristics of the two types of compounds have been classified by Semour, Ref 1.

TABLE III.
CHARACTERISTICS OF ORGANIC AND INORGANIC COMPOUNDS

ORGANIC	INORGANIC
1. All contain carbon.	1. Only a few contain carbon. e.g.; cyanides, carbonates
2. Exist as simple molecules. Some are gases and many are liquids at room temperature. Most solids melt at low temperature.	2. Exist as ionic crystals and as ions in water solutions. Many solids have an extremely high melting point.
3. Most molecules contain carbon bonds that are very stable.	3. Ions of molecules containing like atoms bonded together are uncommon and usually are unstable.
4. Undergo combustion in air readily.	4. Combustion in air is uncommon.
5. Are usually soluble in some other organic compounds, but only a few of those with low molecular weight or with special structure are soluble in water.	5. Solubility in water more common than solubility in organic compounds.
6. Isomerism, the existance of two or more substances with the same molecular weights, is common.	6. Isomerism is somewhat rare.
7. Probably between 500,000 and 1,000,000 compounds are known.	7. Probably less than 100,000 compounds are known.

CLASSIFICATION OF ORGANIC COMPOUNDS

Structures which have the same general formula belong to a
__homologous__ __series.__

ALKANES

Hydrocarbons having the general formula C_nH_{2n+2}
are called __paraffin, saturated__ or __alkane.__

ALKANE HYDROCARBONS

Methane	CH_4	Octane	C_8H_{18}
Ethane	C_2H_6	Nonane	C_9H_{20}
Propane	C_3H_8	Decane	$C_{10}H_{22}$
Butane	C_4H_{10}	Hendecane	$C_{11}H_{24}$
Pentane	C_5H_{12}	Dodecane	$C_{12}H_{26}$
Hexane	C_6H_{18}		

ALKYLS

Alkyl radicals are single valent and are those derived
from other more complex compounds. Alkylization consists
of adding such radicals to other compounds. Alkylization of
crude oil improves the fluidity and upgrades the product.

STRUCTURE OF ALKYL RADICALS

CH_3- CH_3-CH_2-- $CH_3-CH_2-CH_2-$

__Methyl__   __Ethyl__   __n-Propyl__

(n for normal)

$$CH_3 - \overset{\overset{H}{|}}{\underset{\underset{CH_3}{|}}{C}} -$$

$CH_3-CH_2-CH_2-CH_2-$

$CH_3-CH_2-\overset{}{\underset{|}{CH}} - CH_3$

__n-Butyl__   __s-Butyl__

(s for secondary)

__Isopropyl__

$$CH_3 - \overset{\overset{H}{|}}{\underset{\underset{CH_3}{|}}{C}} - CH_2-$$

$$CH_3 - \overset{\overset{CH_3}{|}}{\underset{\underset{CH_3}{|}}{C}} -$$

__Isobutyl__   __t-Butyl__

7   (t for tertiary)

ISOMERS

Isomer is the name given by Berzelius to a compound having the same molecular weight but with different structural arrangement. For instance, consider the isomer of butane, C_4H_{10}.

```
  H H H H
  | | | |
H-C-C-C-C-H
  | | | |
  H H H H

  n-Butane
```

```
  H H H
  | | |
H-C-C-C-H
  | | |
  H 1 H      note that the
    |        methyl radical
  H-C-H      has been substituted
    |        for one hydrogen
    H        atom
```

UNSATURATED ALKANES (ALKENES)

Unsaturated alkanes are known as olefins or alkenes. These have the formula C_nH_{2n}. They can become saturated by treatment with surfuric acid, H_2SO_4. The list includes ethylene and propylene.

ALIPHATIC COMPOUNDS

Aliphatic compounds are those organic compounds which contain open chains of carbon atoms, as opposed to the closed rings of carbon atoms of the aromatic compounds.

BENZENE AND AROMATIC COMPOUNDS

Benzene was discovered in 1825 by Farady, the English chemist. It has the formula C_6H_6. It was investigated by Kekule' von Stradonitz (1829-1896). Kekule' developed the ring theory of benzene which assumes a resonant condition of alternating bonds between the carbon atoms. One of Kekule's major contributions was his use of diagrams to show the structure of the molecule.

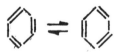

Aromatics are based on the benzene ring. The term aromatic in chemistry refers to the ring structure, although it originated in the false supposition that a fragrant aroma was related to structure. Important aromatic hydrocarbons include benzene, hexane, cyclohexane, toluene and xylene. Most artificial dyestuffs contain one or more benzene rings.

ALCOHOLS AND ETHERS

The -OH, or hydroxyl, radical is common to all alcohols. Common alcohols are:

Methanol (Methyl Alcohol)	CH_3OH
Ethanol (Ethyl Alcohol)	C_2H_5OH
Propanol	C_3H_7OH
Pentanol	$C_5H_{11}OH$
Hexanol	$C_6H_{13}OH$

Ethers and alcohols have the same general formula, $C_nH_{2n+2}O$. Ethers differ in structure in that the oxygen atom is bonded to a carbon and a hydrogen atom. For instance, we have dimethyl ether, CH_3OCH_3.

ESTERS

Esters are the product of an alcohol and an acid. All of the low molecular weight esters are liquid with a pleasant fruity odor. For instance, amyl acetate has the odor of bananas. Esters are used in lacquers and plastics.

ALIPHATIC ACIDS

Some alcohols can be converted to vinegar by exposure to air, forming acetic acid, an alaphatic acid.

An aliphatic acid has the radical -COOH in its structure, written diagramically as

$$-\overset{}{\underset{\overset{\|}{O}}{C}}-OH$$

Simple aliphatic acids can be classified according to the number of carbon atoms:

1 Carbon	Formic Acid
2	Acetic Acid
3	Propionic Acid
4	Butyric Acid
5	Valeric Acid
12	Lauric Acid
18	Stearic Acid

ALDEHYDES AND KETONES

The amount of hydrogen in alcohol can be reduced to form aldehydes and ketones. The product obtained from primary alcohols are aldehydes; the product obtained from secondary alcohols are ketones. Ketones are found to be good solvents.

ALDEHYDES	KETONES
Formaldehyde	Acetone
Acetaldehyde	Methyl ethyl ketone
Propionaldehyde	Methyl propyl ketone
Butyraldehyde	Diethyl ketone

ALKYL HALIDES

The halagons are composed of Fluorine, Fl, Chlorine, Cl, Bromine, Br, Iodine, I, and Astatine, At. These elements form Group VII of the Periodic Table. The term means "salt former." Halogens can be reacted with hydrocarbons to form alkyl halides, where one of the halogens replaces one of the hydrogen atoms.

NOMENCLATURE OF THE HALIDES

Chloromethane	(Methyl Chloride)	CH_3Cl
Dichloromethane	(Methylene Chloride)	CH_2Cl_2
Trichloromethane	(Chloroform)	$CHCl_3$
Tetrachloromethane	(Carbon Tetrachloride)	CCl_4
Chloroethane	(Ethyl Chloride)	C_2H_5Cl
1-Bromobutane	(n-Butyl Bromide)	$C_3H_7CH_2Br$

AMINES AND NITROGEN COMPOUNDS

Amines are compounds formed by replacing hydrogen atoms of ammonia, NH_3, with organic radicals. They are classified into primary amine types as

NH_2R, NHR_2, and NR_3 where R is the organic radical.

SIMPLE AMINES

Methylamine	CH_3NH_2
Dimethylamine	$(CH_3)_2NH$
Trimethylamine	$(CH_3)_3N$
Triethylamine	$(C_2H_5)_3N$

COMPRESSIBILITY

The ideal gas law of Boyle and Charles describes specific weight as being proportional to the temperature and pressure. This is not quite correct where molecules are in close contact as near the critical point and at supercritical conditions.

The reasons for the deviation is that (1) the gas molecules are not point dimensions but occupy finite diameters, and (2) at very close distances the gas molecules are subject to attractive forces.

Non-ideal gas states have been described by a great many equations. One of the most widely used is the van der Waals Equation, for which experimental coefficients are found in many reference books.

Another approach is that of Pitzer and associates. In a paper published in 1955, he devised a three-part graphical solution. Part One consists of determining a compressibility factor for a "simple" fluid. Part Two is to determine a correction factor for deviation from the simple fluid. Part Three is determination of an "acentric factor" for the particular substance. The compressibility factor, which is applied to the ideal gas law equation, is then found by adding the product of the the two correction factors to the factor for the simple fluid. This method of correction is amenable to computer analysis.

Part One: Correction Factor for Simple Fluid

This correlation plots the value of Z_o, the correction factor, versus the logarithm of the reduced pressure, P_r (the ratio of the pressure to the critical pressure). There are designated lines of constant reduced temperature, T_r (the ratio of the temperature to the critical temperature).

The envelope of lines shows a lower limit for the liquid state. This line is found to be nearly linear for reduced pressure vs. correction factor.

$$Z_L = J \, P_r^{\,n}$$

where Z_L = Simple-Fluid Correction Factor for liquid
J = 0.1742
n = 0.8546
P_r = Reduced Pressure

The envelope of the Pitzer Chart shows for the vapor a Z_o approaching 1.0 for values of reduced pressure and reduced temperature of <1. This is to be expected and indicated that the perfect gas law is applicable at these values. For a reduced temperature of 1.0, there is essentially a parabolic reduction in the correction factor up to the reduced pressure value of 1.0, at which the vapor and liquid values essentially converge. At higher values of the reduced temperature, there is less reduction and convergence to the liquid line at higher reduced pressures.

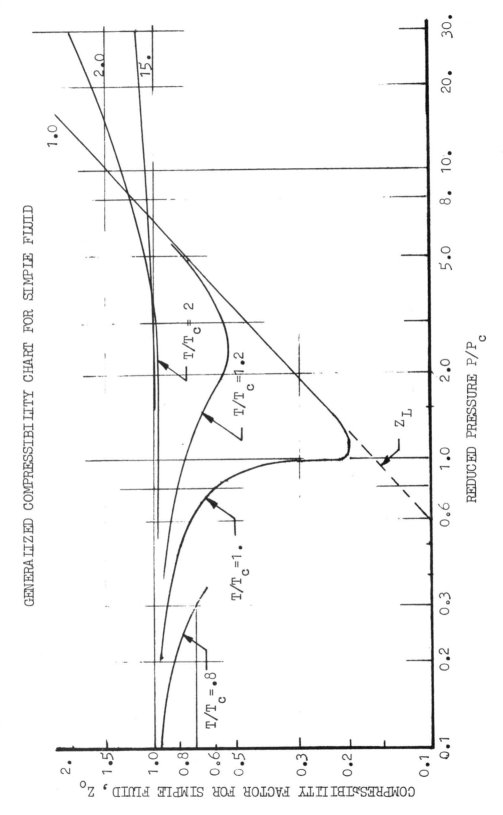

GENERALIZED COMPRESSIBILITY CHART FOR SIMPLE FLUID

12

The most practical approach to mathematical representation
of the graph appeared to be to determine the correction factor of
a simple fluid as the sum of the liquid line plus an increment
for the vapor line.

The latter was determined by the use of fourth order determinants. Six sets of values were found, each set being applicable
to a particular T_r value from 1.0 to 4.0. Interpolation can be
made for intermediate T_r values.

TABLE IV.
COEFFICIENTS FOR COMPRESSIBILITY FACTOR
OF SIMPLE FLUID

T_r	A	B	C	D	P_r Limits
1.0	0.7653	-4.0547	3.1614	-0.7234	0 to 2.0
1.15	-0.171	-0.414	0.9190	-0.022	0 to 3.0
1.30	-0.2693	-0.0784	0.0348	-0.0031	0 to 5.0
1.60	-0.2536	0.0149	0.0006	0.	0 to 6.0
2.0	-0.2178	0.0194	-0.0009	0.	0 to 7.0
4.0	-0.1631	0.0100	-0.0005	0.	0 to 8.0

$$dZ_v = 1 + A\,P_r + B\,P_r^2 + C\,P_r^3 + D\,P_r^4$$

$$Z_o = Z_L + dZ_v$$

Part Two: Correction to Compressibility of Simple Fluid

This correction factor of the Pitzer Chart is plotted
versus the logarithm of reduced temperature for lines of constant
reduced pressure. The correction is nearly zero at values of
1.0 for values of 1.0 reduced pressure and reduced temperature.
At a reduced pressure of 9.0, the correction is nearly -0.040 at
reduced temperature of 0.8 and is nearly 0.50 at reduced temperature of 3.0.

In general, the correction curves at constant reduced pressure exhibit a sine wave shape but with considerable irregularity,
sufficient to suggest imperfect correlation. Accordingly, these
curves have been represented by simple two-part straight lines.

TABLE V.
COEFFICIENTS FOR CORRECTION TO COMPRESSIBILITY
OF SIMPLE FLUID

P_r	A	B	Transition T_r	A'	B'
1.0	-0.303	0.2917	0.9	0.0137	0.0041
2.0	-3.70	3.00	1.3	0.258	-0.0444
5.0	-0.653	0.492	2.0	0.460	0.065

$$Z' = A + B\,T_r \quad \underline{or} \quad = A' + B'\,T_r$$

Part Three: Acentric Factor for Fluid Compressibility

The acentric factor can be considered a type of vapor pressure correlation. It is determined by Pitzer as:

$$w - -(\log P_r + 1.000) \text{ at } T_r = 0.7$$

A second correlation is given by Edmister, Ref. 2.

$$w = 3/7 \ (\log P_c/((T_c/T_b)-1))- 1$$

where P_c = Critical Pressure,   atmospheres
T_c = Critical Temperature.
T_b = Normal Boiling Point (1 atm.)

COMPOSITE COMPRESSIBILITY CORRECTION FACTOR

$$Z = Z_o + w \ Z'$$

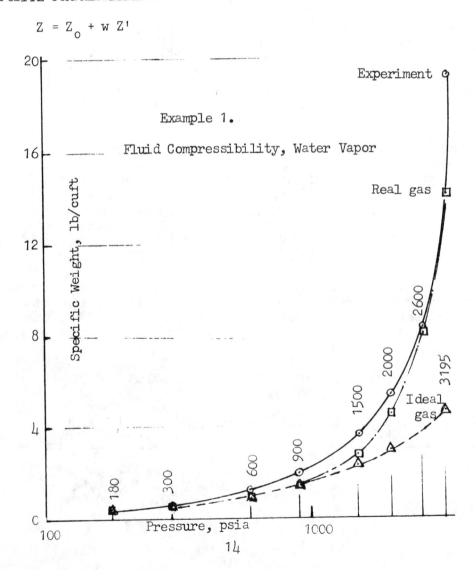

Example 1.

Fluid Compressibility, Water Vapor

Example 1. Fluid Compressibility, Water Vapor
(See BASIC Computer Program # 1)

Data: Critical Temperature, T_c = 1161 R (706 K)
       Critical Pressure, P_c = 3195 psia (217.5 atm)
       Normal Boiling Temp., T_{nb} = 212 F (672 R)

Object: Evaluate specific weight at 2600 psia and 674 F.

Procedure: T_r = (674+460)/1161 = 0.976

             P_r = 2600/3195   = 0.8138

1. Determine specific weight for perfect gas

$$g\rho = P/(RT)$$

$$= (2600 \times 144)/((1545/18.0) \times (674+460))$$
$$= 3.85 \text{ lb/cuft}$$

2. Determine correction factor for simple fluid

2a. Find liquid factor

$$Z_L = 0.1742 \, (0.976)^{0.8546} = 0.1706$$

2b. Find vapor increment factor, use coefficients for T_r=1.

$$dZ_v = 1 + 0.7653(0.8138) + (-4.0547(0.8138)^2$$

$$+ 3.1614(0.8138)^3 - 0.7234(0.8138)^4$$

$$= 0.324$$

2c. Find $Z_o = Z_L + dZ_v$

$$= 0.1706 + 0.324 = 0.4946$$

3. Find correction for non-simple fluid

Use values for P_r = 1.0; T_r more than 0.9

$$Z' = 0.0137 + (0.0041 \times 0.976) = 0.0177$$

4. Find acentric factor

$$w = 3/7(\log 219.5/((1161/212)-1))-1 = -0.78$$

5. Compute compressibility factor
$$Z = 0.4946 + (-0.78 \times 0.0177) = 0.481$$
6. Compute real gas specific weight
$$g\rho = 3.85/0.481 = 7.99 \text{ lb/cuft}$$

7. Fugacity press = $7.99(1545/18)(674+460)/144$ = 5400 psia

Example 1.   SPECIFIC WEIGHT FOR REAL GAS

Values for T_r = 1.0   Water Vapor

TABULATION OF RESULTS, COMPRESSIBILITY CORRECTION
Acentric Factor
w = 0.379

Pressure	Reduced Pressure	Temp.	Reduced Temp.	Simple Fluid	Corr. Simple Fluid	Compress. Factor
P psia	P_r	T °F	T_r	Z_o	Z'	Z
180	0.056	373	0.717	1.046	-0.094	1.010
300	0.094	417	0.755	1.062	-0.083	1.031
600	0.188	486	0.815	1.062	-0.065	1.037
900	0.282	532	0.854	1.019	-0.054	0.999
1500	0.469	596	0.910	0.849	0.017	0.855
2000	0.626	636	0.944	0.671	0.018	0.678
2600	0.814	674	0.977	0.470	0.018	0.477
3195	1.000	706	1.000	0.323	0.018	0.330

TABULATION OF RESULTS, SPECIFIC WEIGHT VALIDITY CHECK

Pressure psia	Temperature deg F	Specific Weight, lb/cuft		
		Perfect Gas	Corrected	Kent's Table
180	373	0.363	0.359	0.395
300	417	0.574	0.557	0.649
600	486	1.065	1.027	1.303
900	532	1.523	1.525	2.012
1500	596	2.385	2.789	3.648
2000	636	3.064	4.519	5.333
2600	674	3.850	8.071	8.299
3195	706	4.646	14.079	19.157

16

BASES FOR COMPUTATION

In scientific work, particularly chemistry, the CGS system of measurement is most often used. Caution is needed for constants which may be system oriented.

The fundamental reference is the Standard Reference State at zero degrees Kelvin and a pressure of one atmosphere.with the substance in the crystalline form.

The Standard Reference Temperature is 25 deg C (537 deg R) and the Standard Pressure is one atmosphere (14.69 psia).

Thermodynamic properties of substances are prepared for a variety of base temperatures and conditions according to the expected uses of the tables. For instance, mercury vapor tables prepared by General Electric Co. have a base of 32 deg F; tables of saturated ammonia prepared by the Bureau of Standards hava a base of -40 deg F.

Non-reacting systems, such as use of steam for driving a turbine, need not consider the energy of formation of the substance. Only property change with pressure and temperature need be considered.

Reacting systems, such as combustion, dissociation, or the formation or dissolution of chemicals, must consider not only property change with pressure and temperature but the energy of formation of the substance.

Total enthalpy, H, must include energy of formation plus the energy change from the Standard Reference State to the base of computation.

$$H = dH_{0-537R} + dH_{537-T} + H_{formation}$$

INTRINSIC THERMODYNAMIC PROPERTIES

The three intrinsic thermodynamic properties essential for chemical process analysis are specific heat, heat of combustion, and heat of formation.

Specific heat can be found experimentally by adding a given mass of a substance to a second given mass of water at a different temperature in an insulated container. From a measurement of the final temperature the specific heat of the first substance can be found. (The specific heat of water is by definition equal to 1.0.)

The heat of combustion can be found in a similar manner. The substance is burned in air in an insulated water filled jacket. The increase in temperature determines the heat of combustion.

Heat of combustion can be predicted by a knowledge of the strength of the chemical bonds. Such a method appears to have been first developed by H. Stanley Redgrove in 1909 following publication in 1886 by Julius Thomsen, a German chemist, of experiments with a large number of combusted substances. It is informative and instructive to consider in some detail the Redgrove methods, which apply also to heat of formation, Ref.3,4,5.

A more recent, but still an essentially pioneer work was that of T. L. Cottrell, The Strength of Chemical Bonds, which was published in 1954, Ref. 6.

Specific heat correlations of great value include the work of the American Petroleum Institute Research Project 44 (API RP 44) as published by the U.S. Department of Commerce, Selected Values of Properties of Hydrocarbons, 1947. A more general, and more recent work is the JANAF Thermochemical Tables, 1970 (Originally, Joint Army Navy Air Force) also published by the U.S. Department of Commerce, Ref. 7 and 8 , respectively.

SPECIFIC HEAT

The change of specific heat with temperature change has been related to step increases in molecular energy levels. The specific heat of most substances can be represented by a four part set of constants for temperatures above 540 deg R. Hydrogen is an exception, requiring a five part set of constants because of its greater variations.

Values for specific heat coefficients are given in Table 6.

Specific heat is usually tabulated in terms of constant pressure, c_p, rather than constant volume, c_v. The two are related by the constant R.

$$c_v = c_p - 1.986 \text{ BTU/mol-}^\circ R$$

USE OF DETERMINANTS TO FIND COEFFICIENTS
FOR SPECIFIC HEAT EQUATION

Determinants is the name given to a set of numbers in a square array which represent the coefficients of a set of equations. The number of equations must equal the number of unknowns.

Determinants can be used to evaluate an expression for specific heat, based on experimental data. The equation is an exponential expansion:

$$C_p = C_1 + C_2 T + C_3 T^2 + C_4 T^3 + C_5 T^4$$

where T is the temperature

For most substances, the variation of specific heat over the usable range of temperatures is small enough that the equation can be reduced to four constants. That is, the constant C_5 is equal to zero. Hydrogen is an exception.

Specific heat approaches zero at zero temperature. At zero temperature, then, $C_p = C_1$ as graphically determined. The set of numbers for the C_p determinant can then be reduced to become:

$$C_p - C_1 = C_2 T + C_3 T^2 + C_4 T^3$$

Solution requires first that the value of the base determinant be found. This is found by

$$
D_o = \begin{vmatrix} T_1 & T_1^2 & T_1^3 \\ T_2 & T_2^2 & T_2^3 \\ T_3 & T_3^2 & T_3^3 \end{vmatrix}
\begin{aligned}
&= (T_1\, T_2^2\, T_3^3) + (T_2\, T_3^2\, T_1^3) \\
&\quad + (T_3\, T_1^2\, T_2^3) - (T_1^3\, T_2^2\, T_3) \\
&\quad - (T_1^2\, T_2\, T_3^3) - (T_1\, T_3^2\, T_2^3)
\end{aligned}
$$

Observe that the multiplication of elements is carried out diagonally, so that each is accounted for.

Values of C_2, C_3 and C_4 are found by substituting in the determinant, the values for the left side of the equation. For example, the determinant numerator of C_2 is

$$
D_{C_2} = \begin{vmatrix} T_1 & (C_{p1} - C_1) & T_1^3 \\ T_2 & (C_{p2} - C_1) & T_2^3 \\ T_3 & (C_{p3} - C_1) & T_3^3 \end{vmatrix}
\qquad \text{then,} \quad C_2 = D_{C_2}/D_o
$$

19

Fourth order determinants can be reduced to third order determinants by taking the elements of the first column, with signs alternately plus and minus, and forming the sum of the products by multiplying each of the elements by its corresponding <u>minor</u>.

$$D = \begin{vmatrix} a_1 & b_1 & c_1 & d_1 \\ a_2 & b_2 & c_2 & d_2 \\ a_3 & b_3 & c_3 & d_3 \\ a_4 & b_4 & c_4 & d_4 \end{vmatrix} = a_1 \begin{vmatrix} b_2 & c_2 & d_2 \\ b_3 & c_3 & d_3 \\ b_4 & c_4 & d_4 \end{vmatrix} - a_2 \begin{vmatrix} b_1 & c_1 & d_1 \\ b_3 & c_3 & d_3 \\ b_4 & c_4 & d_4 \end{vmatrix}$$

$$+ a_3 \begin{vmatrix} b_1 & c_1 & d_1 \\ b_2 & c_2 & d_2 \\ b_4 & c_4 & d_4 \end{vmatrix} - a_4 \begin{vmatrix} b_1 & c_1 & d_1 \\ b_2 & c_2 & d_2 \\ b_3 & c_3 & d_3 \end{vmatrix}$$

Note that the use of determinants to define a curve has been described as pinning a snake at particular points; the deviation at other points may be extreme. This is especially to be checked if the experimental curve has two inflection points, that is, a maximum and a minimum where the slope changes.

Example 2. COEFFICIENTS FOR HYDROGEN
SPECIFIC HEAT
(See BASIC Computer Program # 2)
Data: From "Selected Values of Properties of Hydrocarbons"

$$\text{Temperature, } \begin{array}{l} 540 \text{ R} \\ 1980 \\ 3600 \\ 4950 \\ 0 \end{array} \quad c_p = \begin{array}{l} 3.420 \text{ BTU/lb-deg R} \\ 3.625 \\ 4.055 \\ 4.300 \\ 2.200 \end{array}$$

Procedure:

1. Write the four equations:

$$3.420 - 2.200 = 1.220 = a_1 \, 540 + b_1(540)^2 + c_1(540)^3 + d_1(540)^4$$

$$3.625 - 2.200 = 1.425 = a_2 \, 1980 + b_2(1980)^2$$
$$+ \, c_2(1980)^3 + d_2 \, (1980)^4$$

$$4.055 - 2.200 = 1.855 = a_3 \, 3600 + b_3 \, (3600)^2$$
$$+ \, c_3 \, (3600)^3 + d_3 \, (3600)^4$$

$$4.300 - 2.200 = 2.100 = a_4 \, 4950 + b_4 \, (4950)^2$$
$$+ \, c_4 \, (4950)^3 + d_4 \, (4950)^4$$

2. Use Program 2 to find the third order determinants as the minors and multiply by the first column term not included, then sum with alternating signs:

$$D_o = 540(8.063 \times 10^{30}) - 1980(1.6869 \times 10^{30})$$
$$+ \, 3600(5.2831 \times 10^{29}) - 4950(1.0576 \times 10^{29})$$
$$= 2.3924 \times 10^{33}$$

$$D_a = 1.220(8.0863 \times 10^{30}) - 1.425(1.6869 \times 10^{30})$$
$$+ \, 1.888(5.283 \times 10^{29}) - 2.100(1.0576 \times 10^{29})$$
$$= 8.2368 \times 10^{30} \qquad a = 3.4429 \times 10^{-3}$$

$$D_b = 540(6.9312 \times 10^{24}) - 1980(9.2110 \times 10^{24})$$
$$+ \, 3600(3.2950 \times 10^{24}) - 4950(6.8701 \times 10^{23})$$
$$= -6.0336 \times 10^{27} \qquad b = -2.5220 \times 10^{-6}$$

21

(Example 2, continued)

$$D_c = 540(-2.5201 \times 10^{21}) - 1980(-4.2514 \times 10^{21})$$
$$+ 3600(-2.1743 \times 10^{21}) - 4950(-4.9980 \times 10^{20})$$
$$= 1.7034 \times 10^{24} \qquad c = 7.1202 \times 10^{-10}$$

$$D_d = 540(2.5452 \times 10^{17}) - 1980(4.8884 \times 10^{17})$$
$$+ 3600(3.0662 \times 10^{17}) - 4950(8.6992 \times 10^{16})$$
$$= -1.5724 \times 10^{20} \qquad d = -6.5725 \times 10^{-14}$$

3. Write the equation:

$$c_p = 2.2000 + 3.4429 \times 10^{-3} \, T - 2.5220 \times 10^{-6} \, T^2$$
$$+ 7.1202 \times 10^{-10} \, T^3 - 6.5725 \times 10^{-14} \, T^4$$

4. Check the validity of the calculation by comparison to the initial values.

Temp.	Data	Calc.
540 R	3.420	3.430
1980	3.625	3.647
3600	4.055	4.090
4950	4.300	4.347

TABLE VI. SPECIFIC HEAT AT CONSTANT PRESSURE

$$C_p = C_1 + C_2 T + C_3 T^2 + C_4 T^3 + C_5 T^4$$

where C_p is BTU/lb-deg R and T is deg R

SUBSTANCE	C_1	C_2	C_3	C_4	C_5	Validity
Hydrogen, H_2, Gas	2.2000	3.4429×10^{-3}	-2.5220×10^{-6}	7.1202×10^{-10}	-6.5725×10^{-14}	540R – 4950R
Oxygen, O_2, Gas	0.218	-1.167×10^{-5}	2.743×10^{-8}	-4.439×10^{-12}	0	90R – 4950R
Nitrogen, N_2, Gas	0.248	-1.065×10^{-5}	2.154×10^{-8}	-3.365×10^{-12}	0	=
Nitric Oxide, NO, Gas	0.252	-4.803×10^{-5}	4.453×10^{-8}	-6.656×10^{-12}	0	=
Carbon Monoxide, CO, Gas	0.248	-9.165×10^{-6}	2.233×10^{-8}	-3.577×10^{-12}	0	=
Hydroxyl, OH, Gas	0.353	1.720×10^{-4}	-9.626×10^{-8}	1.375×10^{-11}	0	540R – 4950R
Water, H_2O, Gas 23	0.444	-3.454×10^{-5}	6.952×10^{-8}	-1.030×10^{-11}	0	=
Carbon Dioxide, CO_2, Gas	0.136	1.394×10^{-4}	-3.320×10^{-8}	2.692×10^{-12}	0	=
Carbon, C, Solid	0.033	2.417×10^{-4}	4.767×10^{-8}	-2.810×10^{-11}	0	540R – 2700R
Oxygen, O, Gas	0.327	-1.375×10^{-5}	3.434×10^{-9}	-2.358×10^{-13}	0	540R – 7200R
Hydrogen, H, Gas	4.930	0	0	0	0	=
Nitrogen, N, Gas	0.355	7.870×10^{-7}	-6.559×10^{-10}	1.215×10^{-13}	0	=
Carbon, C, Gas	0.415	-1.204×10^{-6}	2.315×10^{-10}	7.144×10^{-13}	0	=
Ammonia, NH_3, Gas	0.412	1.8126×10^{-4}	2.6235×10^{-8}	-7.8169×10^{-12}	0	540R – 2700R

HEAT OF COMBUSTION

As theorized by Clausius and later proved, only ions, which are dissociated molecules, can take place in reactions. Combustion is a rapid self-sustaining reaction between a fuel and an oxidizer initiated by attaining some particular kindling temperature. The heat of combustion is ameasure of the strength of the chemical bonds of the ionized particles.

Considerable work was done experimentally at the end of the last century to determine thermodynamic properties of many chemical compounds and elements. Determinations of our present era can be expected to have greater accuracy because of improved purity of products, improved instruments, and greater insight into the structure of matter. Nevertheless, an understanding of the nature and meaning of chemical bonds can be given by study of two of the works of this peridd.

Firstly, Julius Thomsen, a German chemist, published in 1886 the results of his experiments measuring the heat of combustion of 15 hydrocarbons, halogen compounds (14 chlorides, 4 bromides, and 2 iodides), oxygen compounds (8 ethers, 12 alcohols, 3 aldehydes, 2 ketones, 9 esters, and 4 acids), sulfur compounds (including sulphides and mercaptans), and various nitrogen compounds. All data were obtained at 18°C.

An English translation of Thomsen's work was made in 1908. One year later, H. Stanley Redgrove published a work comparing calculated values, by a method devised by himself, to the experimental work of Thomsen. The agreement was very good and the method was relatively simple. The essentials of his approach will now be described:

METHOD OF REDGROVE FOR CALCULATION OF HEAT OF COMBUSTION

Consider first the carbon/hydrogen molecule. Carbon has a valence of 4; hydrogen has a valence of 1. There are four possible combinations for an elementary molecule:

```
       H                         H H
       1                         1 1
     H-C-H                      H-C-C-H
       1                         1 1
       H                         H H
```

$\underline{\text{Methane}}$, CH_4 $\underline{\text{Ethane}}$, C_2H_6

```
                                 H-C-H
   H-C≡C-H                        11
                                 H-C-H
```

$\underline{\text{Ethene (Ethylene)}}$, C_2H_2 $\underline{\text{Ethyne(Acetylene)}}$, C_2H_2

Observe that, if we consider Hydrogen to have a single bond and Carbon, by itself, to have no bond, then the four compounds have carbon bonds of 0, 1, 2 and 3.

Redgrove's method is to take the experimental value for the heat of the combustion of the simplest molecule (in this case, Methane) and from this value to compute the heat of combustion of the more complex molecules by consideration of the number of carbon and hydrogen atoms and the number of various bonds.

His formula was derived on the basis that the heat of combustion for hydrocarbons is the algebraic sum of the following heats (of which numbers 1,2,4,5,6 and 7 are negative):

1. The heat due to the severance of the Carbon-Hydrogen Link.
2. The heat due to the decomposition of oxygen molecules into free atoms.
3. The heat of formation of molecules of liquid water from hydrogen and oxygen atoms.
4. The heat due to the severance of single Carbon-Carbon links.
5. The heat due to the severance of double Carbon-Carbon links.
6. The heat due to the severance of triple Carbon-Carbon links.
7. The heat due to the decomposition of oxygen molecules into free atoms.
8. The heat of formation of molecules of carbon dioxide from carbon and oxygen atoms.

The coefficients of the heats 1,2 and 3 are always in the ratio 2: 1/2 : 1; therefore, their algebraic sum can be considered one sum.

Likewise, the algebraic sum of heats 7 and 8 can be considered as one constant.

Redgrove developed a general equation which applies to all hydrocarbon molecules:

$$\text{Heat of combustion} = n\alpha - (n + p - a - 1)\beta - a\gamma - b\delta$$

where alpha, $\alpha = 210.8$ Kcal/gm-mole (value for methane)
       beta, $\beta = 52.5$ "   "
       gamma, $\gamma = 89.0$ "   "
       delta, $\delta = 113.0$ "   "
            n = Number of Carbon Atoms
            p = Number of Hydrogen atoms joined to
                non-adjacent carbon atoms
            a = Number of double Carbon bonds
            b = Number of treble Carbon bonds

For example, find the heat of combustion of Ethane, C_2H_6, at 18 deg C.
          (See BASIC Computer Program # 4
     $n = 2$ $a = 0$
     $p = 0$ $b = 0$

H.C. $= 2(210.8) - (2 + 0 - 0 - 1)\ 52.5 - 0 - 0$
      $= 369.1$ Kcal/gm-mole

Redgrove was able to apply his method with some variations to all of the experimental data of Thomsen. Given the known composition of a homologous group, then experimental data on the heat of combustion for four of that group suffice to determine the necessary constants.

HEAT OF FORMATION

Redgrove's method for estimation of the heat of formation of hydrocarbons follows a similar method as for the heat of combustion. The heat of formation of one molecule pertains to the energy absorbed by the formation of a compound from its elements. The heat of formation of a hydrocarbon, from amorphous, molecular carbon and from gaseous, molecular hydrogen is the algebraic sum of the following heats (for which numbers 1 and 2 are negative).

1. The heat due to the decomposition of a sufficient number of carbon molecules to provide free carbon atoms.

2. The heat due to the decomposition of hydrogen molecules into free atoms.

3. The heat of formation of Carbon-Hydrogen links

4. The heat of formation of single Carbon-Carbon links

5. The heat of formation of double Carbon-Carbon links

6. The heat of formation of triple Carbon-Carbon links

Using the same formula as before for heat of combustion, but with new values for the constants, the heat of formation can be estimated for hydrocarbons.

alpha', $\alpha' = 21.2$ Kcal/gm-mole
beta', $\beta' = 15.0$ "
gamma', $\gamma' = 46.0$ "
delta', $\delta' = 89.5$ "

(See BASIC Computer Program # 5)

HEAT OF COMBUSTION at 25C and constant pressure

	Non-condensed	H_2O Condensed
Hydrogen, H_2, gas	51,571 BTU/lb	60,958 BTU/lb
Carbon, C, Solid	14,087	
Carbon Monoxide, CO, g	4,344	
Methane, CH_4, gas	21,502	23,861
Propane, C_3H_8, g	19,929	21,646
n-Pentane, C_5H_{12}, g	19,499	21,072
n-Octane, C_8H_{18}, g	19,256	20,747
Benzene, C_6H_6, g	17,446	18,172

Methane, 23,861 BTU/lb / (1.8(BTU/lb)/(cal/gm))
                 = 13,256 cal/gm
          or, 23,861 x (16 gm/mol/1.8)/ 1000
                 = 212. Kcal/mole

TABLE VII. HEAT OF FORMATION AT 0 R

SUBSTANCE (ideal gas)	MOL WT	HEAT OF FORMATION, H_f^o BTU/lb-mol
Water, H_2O	18.016	-102,788.
Oxygen, O_2	32.000	0
Hydrogen, H_2	2.016	0
Nitrogen, N_2	28.016	0
Carbon Monoxide, CO	28.01	-48,548.
Carbon Dioxide, CO_2	44.01	-169,143.
Carbon, C, (graphite)	12.01	0
Oxygen, O	16.000	105,455.
Hydrogen, H	1.008	92,916.
Nitrogen, N	14.008	153,216.
Ethane, C_2H_6	30.068	-29,731.
Propane, C_3H_8	44.094	-35,068.
n-Octane, C_8H_{18}	114.224	-68,994.

DERIVED THERMODYNAMIC PROPERTIES

The important properties of Internal Energy, Enthalpy and Entropy can be derived from the intrinsic properties of specific heat, heat of combustion and heat of formation. Specific heat is the most important of these, and for many purposes suffices.

INTERNAL ENERGY

Internal energy, E, is composed of the potential energy of the substance's molecular structure, molecular type and the various energy functions of these constituents. According to the application, it may include the heat of formation of the molecule.

Joule's experiment showed that internal energy is independent of the volume and depends upon only the temperature. This experiment allowed a pressurized gas in one container to flow through a porous plug (with no work being done) into a second container. At equilibrium, with the system insulated, the temperature had not changed. The internal energy remained constant.

If one mole of a gas is considered, and volume is constant, we can see that internal energy can be evaluated by the specific heat at constant volume, c_v.

$$dE = c_v \, dT$$

where dT is the change in temperature

$$\text{or} \quad \Delta E = \int_1^2 c_v \, dT$$

The absolute level of internal energy is not readily determinable, but changes in internal energy can by measured by its relation to work and external energy.

$$E = E_2 - E_1$$

$$\text{and} \quad E = Q - W$$

where Q = Heat added or subtracted
W = Work performed

For processes in which no work is done, at constant volume and pressure, the change in internal energy is equal to the heat of the reaction.

$$E = Q$$

Likewise, where there is no heat change, the work is equal to the energy change

$$E = -W$$

ENTHALPY

Enthalpy is the sum of the internal energy and of the external work on the substance.

$$H = E + PV/J$$

For a perfect gas, $PV = RT$

therefore, $H = E + RT/J$

From above, $E = c_v \, dT$

Also, $c_p = c_v + R$ <u>or</u> $c_v = c_p - R$

Therefore, $H = c_p \, dT$ $\qquad E = c_p \, dT - RT/J$

As noted previously, specific heat can be expressed as an exponential series in terms of temperature:

$$c_p = C_1 + C_2 T + C_3 T^2 + C_4 T^3 + C_5 T^4$$

We can then integrate the value of enthalpy by parts

$$dH = \int_1^2 c_p \, dT$$

$$dH = \Big[C_1 T + (1/2) C_2 T^2 + (1/3) C_3 T^3$$

$$+ (1/4) C_4 T^4 + (1/5) C_5 T^5 \Big]_{T_1}^{T_2}$$

The brackets indicate that the equation is to be evaluated at each specified temperature with the two values subtracted. The value of T_1 is usually the standard temperature, 537 deg R.

Absolute enthalpy should include the heat of formation, H_f, plus the enthalpy from the base temperature of zero which may be in two parts:

$$H = H_{0-537} + H_{537-T} + H_f$$

30

Enthalpy can be a measure of the work performed.

$$\text{Work} = W = P_1 V_1 - P_2 V_2$$

$$\text{or} \quad dW = \int_1^2 PV$$

Differentiating the equation partially, once with respect to pressure and again with respect to volume, and keeping the other variable constant:

$$W = \int_1^2 V \, dP + \int_1^2 P \, dV$$

At constant pressure the first term vanishes:

$$dW = P \, dV \quad \text{foot-pounds}$$

From the above section on Internal Energy:

$$E = Q - W \quad \underline{\text{then}} \quad E = Q - P \, dV/J \quad \text{BTU}$$

$$\text{or} \quad Q = E + \frac{P \, dV}{J}, \text{ which is enthalpy}$$

RELATION BETWEEN ENTHALPY AND INTERNAL ENERGY

We have seen, above, that Internal Energy, E, is found by the integral of c_v dT, and Enthalpy is the integral of c_p dT.

The values of c_v and c_p are, moreover, related by the gas constant, R:

$$c_v = c_p - R$$

where R = 1.986 BTU/mole-deg R

Then, the Internal Energy, E is:

$$E = \int c_p \, dT - \int R \, dT$$

$$= \text{Enthalpy} - RT$$

ABSOLUTE OR TOTAL ENTHALPY

Chemical Processing is mostly concerned with temperatures above room temperature because of the faster reaction rates. For this reason, the Standard base temperature is taken as 77 F (537 R) or 25 C (298 K).

Although other systems are used, it appears rational to base computations on the total energy at absolute zero temperature. The variation of specific heat near absolute zero, and other considerations, make it impractical to compute the change in enthalpy between zero and Standard Temperature. For this reason, tabulated values, taken from the JANAF Tables, Ref 8, are given in Table 8, for enthalpy change between 0 and 537 R.

TABLE VIII. ENTHALPY CHANGE, dH FROM 0 R TO 537 R

SUBSTANCE (Ideal Gas)	Molecular Weight	dH_{0-537} BTU/lb-mol-deg R
Water, H_2O	18.016	4260.6
Oxygen, O_2	32.066	3735.
Hydrogen, H_2	2.016	3643.2
Nitrogen, N_2	28.016	3729.6
Carbon Monoxide, CO	28.01	3729.6
Carbon Dioxide, CO_2	44.01	4028.4
Oxygen, O	16.000	2894.4
Hydrogen, H	1.008	2665.8
Nitrogen, N	14.008	2665.8

Example 3. ENTHALPY CHANGE

(See BASIC Computer Program # 6)

Object: Determine enthalpy change for Oxygen, O_2 from 537 R to 1800 R.

Data: From Table 6 , Specific Heat, c_p, coefficients

$$C_1 = 0.218$$
$$C_2^1 = -1.167 \times 10^{-5} \qquad C_4 = -4.439 \times 10^{-12}$$
$$C_3 = 2.743 \times 10^{-8} \qquad C_5 = 0$$

mol. wt = 32.000

Procedure:

1. Find value of enthalpy, H, at 537 R.

$$H = 0.218(537) + (1/2)(-1.167 \times 10^{-5}(537)^2$$
$$+ (1/3)(2.743 \times 10^{-8}) + (1/4)(4.439 \times 10^{-12})$$
$$+ (1/5)(0)(537)^5$$

$$= 116.71 \text{ BTU/lb}$$

2. Similarly, find value of H at 1800 R.

$$H = 415.20 \text{ BTU/lb}$$

3. Find dH per mole

$$dH = (415.20 - 116.71) \times 32.000$$

$$= 9,551.68 \text{ BTU/mole}$$

TABLE IX.   CHANGE OF ENTHALPY, T-537R
HYDROGEN, H_2, gas

mol. wt = 2.016

Temperature	Enthalpy, H BTU/lb	dH, BTU/mole calc.	dH, BTU/mole NBS
537 R (298 K)	1558.	0	0.
600 R (333 K)	1775.	438.	445.
1800 R (1000 K)	6,221.	9,401.	8,896.
3600 R (2000 K)	12,868.	22,802.	22,767.
5000 R (2778 K)	19,378.	35,237.	-
5400 R (3000 K)	20,633.	38,457.	38,092.

TABLE X.   CHANGE OF ENTHALPY, T-537R
WATER, H_2O, gas

mol. wt = 18.016

Temperature	Enthalpy, H BTU/lb	dH, BTU/mole calc.	dH, BTU/mole NBS
537 R	236.82	0.	0.
600 R	264.85	505.	508.
1800 R	851.36	11,071.	11,187.
3600 R	2036.25	32,184.	31,311.
5000 R	3075.54	51,142.	-
5400 R	3353.44	56,149.	54,387.

TABLE XI.   CHANGE OF ENTHALPY, T-537R
OXYGEN, O_2, gas

mol. wt = 32.000

Temperature	Enthalpy, H BTU/lb	dH, BTU/mole calc.	dH, BTU/mole NBS
537 R	116.71	0.	0.
600 R	130.53	442.	445.
1800 R	415.20	9,552.	9,769.
3600 R	949.37	26,654.	25,469.
5000 R	1,393.45	40,855.	-
5400 R	1,503.17	44,367.	42,174.

34

DIMENSIONLESS TOTAL HEAT FUNCTION, D_H

It is convenient to reduce the enthalpy plus the heat of formation to a dimensionless total heat function. This is done on the basis of a reference temperature of zero degrees and one atmosphere, and for energy in terms of BTU/mole.

The enthalpy change and heat of formation are both in terms of BTU/mole so that division by T, in deg R, and $R/J = 1545/778$ equals 1.986 BTU/mol-deg R, gives a dimensionless result.

$$D_H = (dH_{0-537} + dH_{T-537} + H_f o)/(1.986\ T)$$

Example 4. DIMENSIONLESS TOTAL HEAT FUNCTION

OBJECT: Find value of D_H for water vapor, H_2O, at 600 R and 5000 R

Data: Change of Enthalpy, H_2O

$$dH_{600-537} = 505.\ \text{BTU/mole}$$

$$dH_{5000-537} = 51,142.\ \text{BTU/mole}$$

Heat of Formation and dH_{0-537}

$$H_f o = -102.788.\ \text{BTU/mole}$$

$$dH_{0-537} = 4260.6\ \text{BTU/mole}$$

Procedure:

$$D_{H\ 600} = (4260.6 + 505 - 102,788)/(1.986 \times 600)$$

$$= -82.19$$

$$D_{H\ 5000} = (4260.6 + 51,142 - 102,788)/(1.986 \times 5000)$$

$$= -4.7719$$

TABLE XII. DIMENSIONLESS TOTAL HEAT FUNCTION, D_H

Temperature °R	Carbon Dioxide CO_2	Oxygen O_2	Hydrogen H_2	Nitrogen N_2	Ammonia NH_3
537			3.41	3.50	-11.64
600.	-138.08	3.51			
2000.	- 37.20	3.77	3.65	3.66	1.06
3000.	- 22.31	4.07			
4000.	- 14.82	4.35			
5000.	- 10.50	4.49	3.90	4.17	5.52
1000.			3.53	3.52	-4.01

ENTROPY

Entropy is a measure of the state of organization of the crystalline structure by which molecules group themselves together. At absolute temperature of zero, the freedom of movement of the molecules is zero and the entropy is said to be zero. To correct the semantic disturbance, entropy is usually said to be a measure of the disorganization of molecules.

At higher temperature, higher energy levels, the molecules become more mobile: A liquid consists of layers of incompletely fixed molecules. A gas is totally unorganized. At higher temperatures, the plasma state is reached and the molecules themselves dissociate.

Consider the case of change of entropy at constant temperature. This is surely the most difficult abstraction encountered in the study of energy processes.

We note that Internal Energy, E, is a function only of the temperature.

Enthalpy, H, is the sum of the internal energy plus the effect of pressure and volume change (external work).

Consider the constant volume case:

$$H = E + v \, dp/J$$

Entropy is the ratio of the change of energy at constant temperature. Entropy is designated by the symbol S.

$$dS = dE/T + (v \, dp)/T \, J$$

substituting $dE = c_v \, dT$ and $v = RT/P$

$$dS = \int_1^2 (c_v/T) \, dT + \int_1^2 (R/(JP)) \, dp$$

$$\Delta S = c_v \ln (T_2/T_1) + R/J \ln (P_2/P_1)$$

Similarly, for constant pressure:

$$\Delta S = c_p \ln (T_2/T_1) + R/J \ln (V_2/V_1)$$

At constant temperature, the first term in the two above equations vanishes. Note that we can find entropy for any case by considering it to be first constant volume, and second, constant pressure, taking the sum as the answer.

At constant volume and pressure, the second term vanishes.

ENTROPY FUNCTION, $\emptyset_S$

It is customary to tabulate the first term of the entropy equation, that is, $c_p \ln (T_2/T_1)$, under the symbol Phi, $\emptyset$, called the Entropy Function.

The equation can be integrated by parts for the equation set of specific heat given above.

$$\emptyset = \int_1^2 c_p \, dT/T$$

$$\emptyset_S = \int_1^2 (C_1/T) \, dT + C_2 \int_1^2 (T/T) \, dT + C_3 \int_1^2 (T^2/T) \, dT + C_4 \int_1^2 (T^3/T) \, dT$$
$$+ C_5 \int_1^2 (T^4/T) \, dT$$

$$\emptyset_S = \int_1^2 (C_1/T) \, dT + C_2 \int_1^2 dT + C_3 \int_1^2 T \, dT + c_4 \int_1^2 T^2 \, dT + C_5 \int_1^2 T^3 \, dT$$

use the integral form: $\int x^n \, dx = (1/(n+1)) \, x^{(n+1)}$

$$\emptyset = \Big[C_1 \ln T + C_2 T + (1/2) \, C_3 T^2 + (1/3) C_4 T^3 + (1/4) \, C_5 T^4 \Big]_{T_1}^{T_2}$$

where the brackets indicate solution at each temperature with the difference subtracted

ABSOLUTE ENTROPY

Absolute entropy is referenced to absolute zero

$$S = S_{0-537} + S_{537-T}$$

Example 5. ENTROPY FUNCTION CHANGE, $\emptyset$

Object: Determine entropy change for Oxygen, O_2 gas from 537 R to 1800 R

Data: Specific Heat, c_p, coefficients

$C_1 = 0.218$ $C_4 = -4.439 \times 10^{-12}$

$C_2 = -1.167 \times 10^{-5}$ $C_5 = 0$

$C_3 = 2.743 \times 10^{-8}$ Mol. wt = 32.000

Procedure:

1. Find value of Entropy, S, at 537 R.

$$\emptyset = 0.218 \ln 537 + (-1.167 \times 10^{-5}(537) + (1/2)(2.743 \times 10^{-8}$$
$$\times (537)^2 + (1/3)(-4.439 \times 10^{-12})$$
$$+ (1/4)(0)(537)^4$$

$$= 1.368 \text{BTU/lb-deg R}$$

2. Similarly, find value of entropy at 1800 R.

$$\emptyset = 1.649 \text{ BTU/lb-deg R}$$

3. Find dS per mole

$$\emptyset = (1.649 - 1.358) \times 32.000$$

$$= 8.987 \text{ BTU/mole-deg R}$$

TABLE XIII.   CHANGE OF ENTROPY, T-537 R
Hydrogen, H_2, gas

mol. wt = 2.016

Temperature	Entropy, ∅ BTU/lb-deg R	d∅ BTU/mol-deg R
536 R	15.350	0
600	15.734	0.774
1800	19.813	8.998
3600	22.381	14.174
5000	23.825	17.086
5400	24.129	17.698

TABLE XIV.   CHANGE OF ENTROPY, T-537 R
WATER, H_2O, gas

mol.wt = 18.016

Temperature	Entropy, ∅ BTU/lb-deg R	d∅ BTU/mol-deg R
536 R	2.782	0
600	2.831	0.888
1800	3.358	10.385
3600	3.802	18.372
5000	4.049	22.822
5400	4.102	23.786

TABLE XV.   CHANGE OF ENTROPY, T-537 R
OXYGEN, O_2, gas

mol. wt = 32.000

Temperature	Entropy, ∅ BTU/lb-deg R	d∅ BTU/mol-deg R
536 R	1.368	0
600	1.392	0.773
1800	1.649	8.987
3600	1.852	15.483
5000	1.956	18.826
5400	1.977	19.502

TABLE XVI. ABSOLUTE ENTROPY AT STANDARD TEMPERATURE

| Ideal gas state | dS_{0-536R} | BTU/mole-deg F or Cal/mole-deg K |

SUBSTANCE	dS
Oxygen, O_2	49.003
Hydrogen, H_2	31.211
Water, H_2O	45.106
Nitrogen, N_2	45.767
Hydroxyl, OH	43.888
Nitric Oxide, NO	50.339
Carbon, C, solid	1.3609
Carbon Dioxide, CO_2	51.061
Carbon Monoxide, CO	47.301
Methane, CH_4	44.50
Ethane, C_2H_6	54.85
Propane, C_3H_8	64.51
n-Octane, C_8H_{18}	110.82

DIMENSIONLESS TOTAL ENTROPY FUNCTION, D_S

Entropy can be reduced to a dimensionless quantity in the same manner as for enthalpy.

The dimensions of entropy, S, are BTU/deg R-mole, which are the same as for the gas constant, R. Simply add the entropy at Standard conditions, 537 R and 1 atmosphere, S_{0-537}, to the entropy computed for the applicable temperature, S_{T-537}, and divide by 1.986, the value of R

$$D_S = (S_{0-537} + \emptyset_{T-537})/1.986$$

Example 6. DIMENSIONLESS ENTROPY FUNCTION

Object: Find value of D_S for water vapor, H_2O.
          at 600 R and at 5000R

Data: Change of Entropy, H_2O

$d\emptyset_{600-537}$ = 0.888 BTU/mol-deg R

$d\emptyset_{1800-537}$ = 10.385 BTU/mol-deg R

                    Absolute Entropy at Standard Temperature

dS_{537-0} = 45.106 BTU/mol-deg R

Procedure:

1.   Find absolute entropy at values:

S_{600} = 45.106 + 0.888 = 45.994 BTU/mol-deg R

S_{1800} = 45.106 + 10.385 = 55.491 BTU/mol-deg R

2.   Find dimensionless Entropy Function, D_S

$D_{S,600}$ = 45.994/1.986 = 23.159

D_{S1800} = 55.491/1.987 = 27.941

TABLE XVII. DIMENSIONLESS ENTROPY FUNCTION, D_S
Ideal gas

Temperature °R	Water H_2O	Oxygen O_2	Hydrogen H_2
537	22.70	24.66	15.71
600	23.15	25.02	16.10
1800	27.93	29.15	20.22
3600	31.95	32.42	22.79
5000	34.19	34.14	24.25
5400	34.68	34.44	24.55

Thermodynamic tables sometimes contain values for entropy but without corresponding values for enthalpy. There is a simple relation between entropy and enthalpy from which the latter value can be estimated.

$$\text{Entropy function, } \emptyset = \int_{T_1}^{T_2} c_p \, dT/T = c_p \ln T$$

$$\text{Enthalpy function, } dH = \int_{T_1}^{T_2} c_p \, dT = c_p T$$

$$\text{then, } dH = (\emptyset / \ln T) \, T$$

for standard temperature of 298 K and entropy given in cal/mol-deg K

$$dH = dS \, (298/\ln 298) = dS \times 52.31$$

Thermodynamic properties of hydrocarbons and several other important substances are given in Ref. 7 , based on work of the API Research Project 44. Comparison is made below to the values given by API 44 to those calculated by the above expression.

TABLE XVII.   ENTROPY AND ENTHALPY AT 536R for 0 R datum

SUBSTANCE	ENTROPY, S^o cal/deg K-mol	ENTHALPY, H^o, cal/mol calc.	API 44
Water, H_2O, gas	45.11	2630.	2365.1
Hydrogen, H_2, gas	31.21	1638.	2023.81
Hydroxyl, OH, gas	44.88	2347.	2106.2
Oxygen, O_2, gas	49.003	2563.	2069.78
Nitrogen, N_2, gas	45.77	2394.	2072.27
Carbon Dioxide, CO_2, gas	51.06	2671.	2238.11
Carbon Monoxide, CO, gas	47.30	2474.	2072.63
Carbon, C, gas	37.76	1975.	1558.9
Nitric Oxide, NO, gas	50.34	2633.	2194.2
Methane, CH_4, gas	44.50	2328.	2397.
n-Pentane, C_5H_{12} gas	83.27	4356.	5668.

42

GAS EXPANSION

The expansion of a gas against a finite pressure produces work. When the temperature of the gas is maintained constant by the input of heat, the process is called <u>isothermal</u> (constant temperature). When the temperature is allowed to drop by the extraction of heat from the internal energy of the gas, the process is called <u>adiabatic</u> (no-flow).

ISOTHERMAL EXPANSION

The work of isothermal expansion is:

$$W = \int_{V_1}^{V_2} p\, dv$$

For a perfect gas, $PV = RT$

$$\text{or} \qquad P = RT/V$$

Then,

$$W = RT \int_1^2 dv/V$$

Which integrates as:

$$W = R\,T\, \ln\left(V_2/V_1\right)$$

$$= 1.986 \times \text{MOLWT} \times (\deg R) \ln\left(V_2/V_1\right) \text{ BTU/lb}$$

Now, since $P_1 V_1 = P_2 V_2$ then $V_2/V_1 = P_1/P_2$

$$W = R\,T\, \ln\left(P_1/P_2\right)$$

$$= 1.986 \times \text{MOLWT} \times (\deg R) \times \ln\left(P1/P_2\right) \text{ BTU/lb}$$

Isothermal expansion in terms of volume is given by
BASIC Computer Program # 8.

Isothermal expansion in terms of pressure is given by
BASIC Computer Program # 9.

ADIABATIC EXPANSION

For adiabatic expansion, the heat transfer, Q, is zero. The energy is obtained by a reduction in internal energy, $c_v \, dT$, and the work is - p dV/J, then:

$$C_v \, dT = - p \, dV/J$$

By the perfect gas law, $p = RT/V$

then, $C_v \, dT = - RT \, dV/V$

This can be integrated as

$$C_v \ln (T_2/T_1) = - (R/J) \ln (V_2/V_1) \quad \text{per mole}$$

$$= -(R/J) \ln (V_2/V_1)/\text{MOLWT} \quad \text{per pound}$$

Note that $C_v = C_p - (1.986/\text{MOLWT})$ BTU/deg R-lb

The above relation can be solved for final temperature by
   BASIC Computer program # 10.

The final volume can be found by
   BASIC Computer program # 11.

Example 7.   ADIABATIC EXPANSION TEMPERATURE

Initial temperature, T_1 = 520 R
Expansion volume ratio equal to 5 = V_2/V_1
Specific heat at constant pressure, C_p = 0.240
Molecular weight = 28.96

Find final temperature of adiabatic expansion, T_2

PROCEDURE:

1. Find value of specific heat at constant volume, C_v

   $C_v = C_p - (R/J)/\text{Molwt}$

   $R/J = 1544/778 = 1.98$

   $C_v = -.240 - 1.98/28.96 = 0.1714$

2. Substitute in temperature-volume equation.

   $$\ln (T_2/T_1) = -(R/J) \ln (V_2/V_1)/(C_v \times \text{Molwt})$$

   $$= - (1.98) \ln (5)/(0.1714 \times 28.96)$$
   $$= - 0.6439$$
   $$T_2 = T_1 \times \text{EXP} (-0.6439)$$
   $$= 500 \times 0.5252$$
   $$= 273 \text{ R}$$

44

PRESSURE AND VOLUME RELATION FOR ADIABATIC EXPANSION

As before, the relation between work and energy is:

$$W\ C_v\ dT = -\ P\ dV/J \tag{1}$$

Differentiate the perfect gas equation

$$P\ V = W\ RT \tag{2}$$

$$V\ dP + P\ dV = W\ R\ dT$$

or $\quad dT = (\ P\ dV + V\ dP)/WR \tag{3}$

Then substituting Equation (3) in (1):

$$C_v\ (P\ dV + V\ dP)/\ R = -\ P\ dV/V \tag{4}$$

Multiply by R and divide by PV

$$C_v\ (dV/V) + C_v\ (dP/P) = -\ (R/J)\ (dV/V) \tag{5}$$

Collecting terms:

$$-\ ((R/J) + C_v)\ (dV/V) = C_v\ (dP/P) \tag{6}$$

Now, since $\quad C_p = (R/J) + C_v)$ and $\ k = C_p/C_v$

$$-\ k\ (dV/V) = dP/P \tag{7}$$

Noting the minus sign, we can integrate this as:

$$-\ k\ \ln\ (V_1/V_2) = \ln\ (P_2/P_1) \tag{8}$$

or we can say that $\quad (V_1/V_2)^k = (P_2/P_1)$

The final pressure for adiabatic expansion is given by
BASIC Computer Program # 12.

ENTROPY CHANGE AT CONSTANT PRESSURE OR CONSTANT VOLUME

Entropy, S, is a measure of the disorganization of a substance. At zero absolute temperature, entropy is zero; it increases with temperature, change to a less organized phase or by pressure and volume change.

The product of entropy and temperature, $\Delta S \times T$, represents the unavailable energy of a system. The change in this product is equal to the heat absorbed for a reversible reaction.

$$dS \, T = dQ \qquad \text{for reversible reaction} \qquad (1)$$

The product dS T is greater than dQ for a non-reversible reaction (where there are losses by friction or turbulence). (Many practical processes, such as gas compression or expansion, are nearly reversible.)

$$dS \, T > dQ \qquad \text{for non-reversible reaction} \qquad (2)$$

The relations between entropy, temperature and heat, above, are considered to reflect an expression of The Second Law of Thermodynamics:

> No change in a system of bodies that takes place of itself can increase the available energy of the system.

or as stated by Kelvin:

> It is impossible by means of inanimate material agency to derive mechanical effect from any portion of matter by cooling it below the temperature of surrounding objects.

It is apparent that change of entropy governs the maximum conversion of heat to work. The conversion of heat to work is also related by The First Law of Thermodynamics:

> The total energy of a system remains constant and cannot be increased or diminished by any means.

The First Law is expressed by the equation relating Work, W, Heat, Q, and the mechanical conversion of work to heat, J, which is 778 ft-lb/BTU.

$$Q = W/J \qquad (3)$$

and including a change of energy of the system, dE

$$dQ = dE + W/J \qquad (4)$$

46

Where there is no change of kinetic energy, or by elevation difference (the usual case for gases), the work done on a fluid is:

$$W = P \, dV \tag{5}$$

then
$$dQ = dE + P \, dV/J \tag{6}$$

Now, by the definition of Enthalpy, H

$$dH = dE + PV/J$$

or
$$dE = dH - PV/J$$

$$= dH - P \, dV/J - V \, dP/J \tag{7}$$

then combining Equation 6 and 7:

$$dQ = dH - V \, dP/J \tag{8}$$

Returning to Equation 1 and substituting dQ

$$dS \, T = dH - V \, dP/J$$

or
$$dS = dH/T - V \, dP/(JT) \tag{9}$$

but, from the perfect gas law, $PV = nRT$, $V/T = R/P$

then
$$dS = dH/T - (R/J) \, dP/P \tag{10}$$

For entropy change at constant pressure, labelled $\emptyset_p$

$$\emptyset_p = dH/T = \int_{T_1}^{T_2} c_p \, dT/T \tag{11}$$

For entropy change at constant volume, labelled $\emptyset_v$

$$\emptyset_v = dH/T = \int_1^{T_2} c_v \, dT/T, \text{ and since for a perfect gas}$$

$$c_v = c_p - R/J$$

$$\emptyset_v = \emptyset_p - R/J \int_{T_1}^{T_2} dT/T$$

$$= \emptyset_p - R/J \, \ln (T_2/T_1) \tag{12}$$

Then, entropy change at constant pressure is

$$dS = \emptyset_{2p} - \emptyset_{1p} \tag{13}$$

Entropy change at constant volume is

$$dS = \emptyset_{2p} - \emptyset_{1p} - (R/J) \, \ln (T_2/T_1) \tag{14}$$

ENERGY PROCESSES

We can consider the total energy of a substance to consist of a free and an unavailable constituent:

Total Energy = Free Energy + Unavailable Energy

The Unavailable Energy is the waste heat of the system; the energy flow that does not change the internal energy or produce work. It is the ratio of heat flow, dQ, at temperature T which measures entropy change, dS.

$$dQ/T = dS \quad \underline{or} \quad dQ = T\,dS$$

The Total Energy and the Free Energy are related and have definitions and values depending on the energy processes which are considered.

HELMHOLTZ ENERGY

For a non-flow system, working within fixed boundaries, the energy received, dQ, equals the change in internal energy, dE, plus the work, dW:

$$dQ = dE + dW \quad \underline{or} \quad dW = dQ - dE$$

For a reversible process (no energy loss),

$$dQ = T\,dS$$

Then, $\quad dW \leqq T\,dS - dE$ (the symbol indicates equal or less)

By summing between States 1 and 2,

$$W = T(S_2 - S_1) - (E_2 - E_1)$$
$$= (E_1 - T\,S_1) - (E_2 - T\,S_2)$$

The combination $E - TS$ is a derived property, given the symbol A and known as the Helmholtz function, or the <u>work function</u>.

$$A = E - T\,S$$

GIBBS ENERGY

If a system does work upon its boundaries so that the boundary is no longer fixed, the work accomplished can be measured by the displacement. The work done by the system is the product of the pressure and the change in volume, dV:

$$W = P (V_2 - V_1)/J$$

By reference to the above defined work function, the Helmholtz function, which considers fixed boundaries, the net work becomes:

$$W = (E_1 - T S_1) - (E_2 - T S_2) - P (V_2 - V_1)/J$$

or
$$W = (E_1 + PV_1/J) - (E_2 + PV_2/J) - TS_1 + TS_2$$

but
$$E + PV/J = H, \text{ the enthalpy}$$

then,
$$W = H_1 - H_2 - TS_1 + TS_2$$

This final relationship is called the Gibbs function, or the free energy function, G:

$$G = H - TS$$

Considering the proper direction for work and energy exchange with the environment, the maximum work is

$$W_{max} = - \Delta G = - \Delta H + T \Delta S$$

The negative of the change in the Gibbs function is a measure of the maximum possible work for a system undergoing a change in pressure and temperature. It is the minimum work that must be done of a system to bring about a pressure and temperature process.

CHANGE OF FREE ENERGY AT CONSTANT TEMPERATURE

The Gibbs Free Energy Function and the Helmholtz Work Function are seen to be related:

Helmholtz: $A = E - TS$ (Eq. 1)

Gibbs: $G = H - TS$ (Eq. 2)

also, $H = E + PV$ (Eq. 3)

then, $G = E + PV - TS$ (Eq. 4)

so: $G = A + PV$ (Eq. 5)

For small changes: $dG = dA + P\,dV + V\,dP$ (Eq. 6)

At constant temperature, the change in internal energy, dE, which Joule showed to be a function of temperature only, is zero. At constant pressure, the only work done is P dV, since V dP=0, which equals A, the work function:

then, $dA = -TS$ and $dA = -P\,dV$ (Eq. 7 and 8)

Substituting now Equation 8 in Equation 6,

$$dG = V\,dP \qquad\qquad\text{(Eq. 9)}$$

Rewriting Equation 9 as a partial derivative to describe the change in Gibbs free energy with change in pressure at constant temperature:

$$(dG/dP)_T = V \qquad\qquad\text{(Eq. 10)}$$

which says that the rate of change of free energy with pressure at constant temperature is equal to the volume.

For a perfect gas, $V = nRT/P$ (Eq. 11)

Substituting Equation 11 in Equation 10:

$$\Delta G = \int_1^2 (nRT/P)\,dP$$

which can be integrated by the form $\int dx/x = \ln x$

$$\Delta G = nRT \ln (P_2/P_1)$$

CHANGE OF FREE ENERGY AT CONSTANT PRESSURE

The rate at which free energy changes with temperature at constant pressure can be obtained by taking the partial derivative of the free energy equation:

$$G = H - TS$$

$$(dG/dT)_P = (dH/dT)_P - T(dS/dT)_P - S$$

but, note that $(dH/dT)_P = c_p$

likewise, $T(dS/dT)_P = c_p$

therefore, $(dG/dT)_P = -S$

furthermore, $-S = (G - H)/T$ from above first equation

then, $(dG/dT)_P = (G - H)/T$ Gibbs-Helmholtz Equation

According to this equation, the difference between the change in free energy and the change in enthalpy for an isothermal process, divided by the absolute temperature, is equal to the rate of change in free energy with temperature, while the pressure is kept constant.

THE VAN'T HOFF EQUILIBRIUM BOX

Van't Hoff postulated an equilibrium box to be used by chemists to study reversible reactions. The box is maintained at constant pressure p_m and all reactions occur at constant temperature, T'. Entry and exit to the box is made through semi-permeable membranes with no change in pressure. The box is of such a large size that addition of one mole of gas does not increase the pressure.

Consider the general reaction

$$aA + bB + \ldots \quad\quad gG + hH + \ldots$$

Suppose that these gases, that is, A, B, ..., are added at partial pressures of

$$P_A' \ , \ P_B', \text{ and } P_G' \ , \ P_H'$$

How large a free energy change is needed to force the gases into the box and to force the products out?

The partial pressures of the gases inside the box are

$$P_A, \ P_B \text{ and } P_G, \ P_H \text{ for leaving gases}$$

The change in free energy for each gas to enter the box and leave can be found by the relation

$$dG = nRT \ \ln (P_2/P_1)$$

These effects can be summarized as:

$$dG = aRT \ \ln(P_A/P_A') + bRT \ \ln(P_B/P_B')$$
$$+ \ gRT \ \ln(P_G/P_G') + hRT \ \ln(P_H/P_H')$$

This is equivalent of the expression

$$dG = -RT \ \ln((P_G^g \ \mathrm{x} \ P_H^h)/(P_A^a \ \mathrm{x} \ P_B^b))$$

$$+ \ RT \ \ln((P_G'^g \ \mathrm{x} \ P_H'^h)/(P_A'^a \ \mathrm{x} \ P_B'^b))$$

As a further simplification

$$dG = - \ RT \ \ln K_p + RT \ \ln Q_p$$

where K_p = Equilibrium constant, for one atmosphere

Q_p = Pressure quotient(the term vanishes at 1 atmo.)

THE VAN'T HOFF EQUATION

At Standard Conditions, 25 C (298 K, 537 R) and 1 atmo.

$$dG = -RT \ln K_p \quad \text{where } K_p \text{ is the } \underline{\text{pressure}}\text{ equilibrium constant}$$

<u>or</u> $dG = dH - T dS$ <u>so that:</u> $-RT \ln K_p = dH - T dS$

The latter can be rearranged as the van't Hoff Equation

$$- \ln K_p = dH/(RT) - dS/R$$

DIMENSIONLESS ENTHALPY AND ENTROPY FUNCTIONS

The elements of the van't Hoff equation, above, can be seen to be in a dimensionless form. This is of great convenience in calculations where a great variety of dimensional systems are encountered.

The enthalpy term, in dimensionless form, is D_H

$$D_H = dH/(RT)$$

The entropy term, in dimensionless form, is D_S

$$D_S = dS/R$$

The free energy term, dG is then

$$dG = D_H \, RT - D_S RT \quad \underline{\text{or}} \quad dG = (D_H - D_S) \, RT$$

Also, $\ln K_p = D_H - D_S$

FUGACITY

The foregoing analysis is based on the perfect gas law, $PV = nRT$, which does not apply for real gases at elevated temperatures and pressures. The analysis results can be used with accuracy, however, by applying the compressibility correction given above. G. N. Lewis, in about 1926, introduced the term "fugacity" to describe a fictious pressure which would replace the ideal gas law pressure to comform to actual measurements of gas density.

CRITERIA FOR SPONTANEOUS AND EQUILIBRIUM PROCESSES

The Gibbs Free Energy Function provides a criterion for determining the possibility of spontaneous and equilibrium processes. For a system that is kept at constant temperature and pressure by heat flow to or from the surroundings:

> If there is a decrease in free energy, the process is possible.

> If there is no change in free energy, the system is in equilibrium.

> If there is an increase in free energy, the process will not take place.

Entropy change is a measure of the potential for processes which are isolated from their surroundings with no external heat flow or work done.

> If there is an increase in entropy, the process is possible.

> If there is no change in entropy, the process is in equilibrium.

> If there is a decrease in entropy, the process will not take place.

Note that there is a close relation between Gibbs free energy G and Entropy, S, as related by Enthalpy, H, and temperature, T.

$$G = H - TS$$

$$S = H/T = \langle(dE + P\,dV)/J)/T \text{ for a reversible process}$$

Example 8. DETERMINATION OF PRESSURE EQUILIBRIUM CONSTANTS BY USE OF GIBBS ENERGY FUNCTIONS

(See BASIC Computer Program 13.)

Formation of Ammonia at One Atmosphere

$$0.5 \ N_2 + 1.5 \ H_2 = NH_3$$

$$
\begin{array}{ll}
NH_3 \ \text{Molwt} = 17.02 \\
N_2 \quad \text{"} \quad = 28.01 \\
H_2 \quad \text{"} \quad = 2.016
\end{array}
$$

PROCEDURE:

1. Find Gibbs Energy Function, G, at several temperatures.

$$- G = S_T - (dH_{T-537})/T$$

BTU/lb-R x Molwt= BTU/mol-R
BTU/lb x Molwt = BTU/mol

Temp. R	$-S_T$ BTU/mol-R	dH_{T-537} BTU/mol	$-G$ BTU/mol-R
Ammonia, NH₃			
537	46.01	0	46.01
1000	+5.91	4440	47.48
2000	+14.21	16622	51.91
5000	+29.39	67211	61.96
Nitrogen, N₂			
537	45.01	0	45.01
1000	+4.37	3258	46.12
2000	+9.57	10812	49.17
5000	+17.70	37699	55.17
Hydrogen, H₂			
537	31.21	0	31.21
1000	+4.51	3377	32.34
2000	+9.75	10866	35.53
5000	+16.96	35112	41.15

2. Find Heat of Formation, H^O

Ammonia, NH_3 = -19,746. BTU/mole at 537 R
Nitrogen, N_2 = 0.
Hydrogen, H_2 = 0.

3. Find values of $dG/T = G + dH/T = - R \ln K_p$ and $\ln K_p$

537R: $dG/T = -46.01 -(-0.5 \times 45.01) - (-1.5 \times 31.21)+(-19746/537)$
= -13.46 $\ln K_p = -(-13.46/1.986) = 6.77$

1000R: $dG/T = -47.48 -(0.5 \times 46.12) - (1.5 \times 32.34)+(-19746/1000)$
= 4.36 $\ln K_p = -(4.36/1.987) = -2.19$

2000R: $dG/T = -51.91 -(0.5 \times 49.17) - (1.5 \times 35.53)+(-19746/2000)$
= 16.10 $\ln K_p = -(16.10/1.986) = -8.10$

5000R: $dG/T = -61.96 -(0.5 \times 55.17) - (1.5 \times 41.15)+(-19746/5000)$
= 23.40 $\ln K_p = -(23.40/1.986) = -11.78$

Example 9. DETERMINATION OF PRESSURE EQUILIBRIUM
CONSTANTS BY USE OF ENTROPY AND ENTHALPY
FUNCTIONS

Formation of Ammonia at One Atmosphere

$$0.5 \ N_2 + 1.5 \ H_2 = NH_3$$

PROCEDURE:

1. Using values from Example 8, find Total Heat Function, D_H

$$D_H = (dH_{0-537} + dH_{T-537} + H_f o)/(1.986 \ T)$$

Temp. R	$H_f o$ BTU/mol	dH_{0-537} BTU/mol	dH_{T-537} BTU/mol	D_H
Ammonia, NH_3				
537	-16,744.	4327.	0	-11.64
1000	(at OR)		4440	- 4.01
2000			16622	1.06
5000			67211	5.52
Nitrogen, N_2				
537	0	3730.	0	3.50
1000			3258	3.52
2000			10812	3.66
5000			37699	4.17
Hydrogen, H_2				
537	0	3643.	0	3.41
1000			3377	3.53
2000			10866	3.65
5000			35112	3.90

2. Using values from Example 8, find Total Entropy Function, D_S

$$D_S = (S_{0-537} + \emptyset_{T-537})/1.986$$

Temp. R	S_{0-537} BTU/mol-R	$\emptyset_{T-537}$ -BTU/mol-R	D_S
Ammonia, NH_3			
537	46.01	0	23.16
1000		5.91	26.13
2000		14.21	30.31
5000		29.39	37.95
Nitrogen, N_2			
537	45.01	0	22.65
1000		4.37	24.85
2000		9.57	27.47
5000		17.70	31.56
Hydrogen, H_2			
537	31.21	0	15.71
1000		4.51	17.98
2000		9.75	20.61
5000		16.96	24.24

Example 9, continued

3. Find natural logarithm of pressure equilibrium constants by summation of entropy functions minus summation of enthalpy functions.

$$\ln K_p = \Sigma D_S - \Sigma D_H$$

537R: $\ln K_p = (23.16 - (0.5 \times 22.65) - (1.5 \times 15.71)) - (-11.64)$
$-(0.5 \times 3.50) - (1.5 \times 3.41))$
$= 6.78$

1000R: $\ln K_p = (26.13 - (0.5 \times 24.85) - (1.5 \times 17.98)) - (-4.01)$
$-(0.5 \times 3.52) - (1.5 \times 3.53))$
$= -2.20$

2000R: $\ln K_p = (30.31 - (0.5 \times 27.47) - (1.5 \times 20.61)) - (1.06$
$-(0.5 \times 3.66) - (1.5 \times 3.65))$
$= -8.10$

5000R: $\ln K_p = (37.95 - (0.5 \times 31.56) - (1.5 \times 24.24)) - (5.52$
$-(0.5 \times 4.17) - (1.5 \times 3.90))$
$= -11.78$

Example 10. DETERMINATION OF PRESSURE EQUILIBRIUM CONSTANTS FOR FORMULA CHANGE BY ENTROPY AND ENTHALPY FUNCTIONS
Formation of Ammonia at One Atmosphere

$$N_2 + 3 H_2 = 2 NH_3$$

Use data and methods of Example 9.

537R: $\ln K_p = (2 \times 23.16 - 22.65 - 3 \times 15.71) - ((- 2 \times 11.64)$
$- 3.50 - 3 \times 3.41)$
$= 13.55$

1000R: $\ln K_p = (2 \times 26.13 - 24.85 - 3 \times 17.98) - ((-2 \times 4.01)$
$-3.52 - 3 \times 3.53)$
$= -4.40$

2000R: $\ln K_p = (2 \times 30.31 - 27.47 - 3 \times 20.61) - (2 \times 1.06$
$-3.66 - 3 \times 3.65)$
$= -16.19$

5000R: $\ln K_p = (2 \times 37.95) - 31.56 - 3 \times 24.24) - (2 \times 5.52$
$-4.17 - 3 \times 3.90)$
$= -23.55$

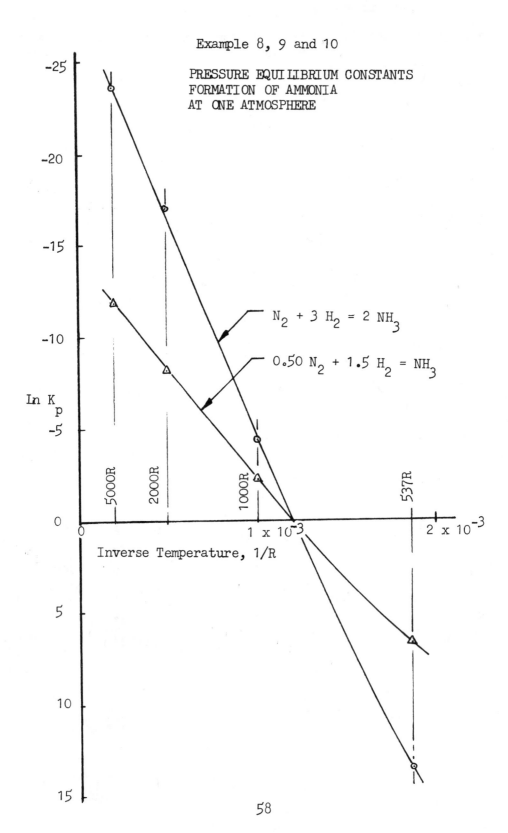

Example 8, 9 and 10

PRESSURE EQUILIBRIUM CONSTANTS
FORMATION OF AMMONIA
AT ONE ATMOSPHERE

$N_2 + 3 H_2 = 2 NH_3$

$0.50 N_2 + 1.5 H_2 = NH_3$

In K_p

Inverse Temperature, $1/R$

1×10^{-3} 2×10^{-3}

5000R 2000R 1000R 537R

-25 -20 -15 -10 -5 0 5 10 15

Example 11. DETERMINATION OF PRESSURE EQUILIBRIUM
CONSTANTS BY USE OF ENTROPY AND ENTHALPY
FUNCTIONS

Dissociation of Water at One Atmosphere

$$2 H_2O = 2 H_2 + O_2$$

H_2O Molwt = 18.016
H_2 " = 2.016
O_2 " = 16.000

PROCEDURE:

1. Find Total Heat Function, $D_H = (dH_{0-537} + dH_{T-537} + H_f o)/(RT)$

Temp. R	$H_f o$ BTU/mol	dH_{0-537} BTU/mol	dH_{T-537} BTU/mol	D_H
Hydrogen, H_2				
537	0	3640.	0	3.41
1800			9401.	3.65
3600			22802.	3.70
5400			38457.	3.93
Oxygen, O_2				
537	0	3723.	0	3.49
1800			9550.	3.71
3600			26654.	4.25
5400			44367.	4.48
Water, H_2O				
537	-102,788.	4254.	0	-92.39
1800			11071.	-24.47
3600			32184.	- 9.28
5400			56149.	- 3.95

2. Find Total Entropy Function, $D_S = (S_{0-537} + \emptyset_{T-537})/R$

Temp. R	S_{0-537} BTU/mol-R	$\emptyset_{T-537}$ BTU/mol-R	D_S
Hydrogen, H_2			
537	31.211	0	18.74
1800		8.96	20.23
3600		14.08	22.80
5400		17.58	24.57
Oxygen, O_2			
537	49.003	0	24.67
1800		8.92	29.16
3600		15.42	32.44
5400		19.44	34.46
Water, H_2O			
537	45.106	0	22.71
1800		10.39	27.94
3600		18.37	31.96
5400		23.79	34.69

59

Example 11, continued

.3.   Find natural logarithms of pressure equilibrium constants.

$$\ln K_p = D_S - D_H$$

537R: $\ln K_p$ = (2 x 18.74 + 24.67 - 2 x 22.71)
              -(2 x 3.41 + 3.49 - (2 x -99.39))
          = -192.36

1800R: $\ln K_p$ = (2 x 20.23 + 29.16 - 2 x 27.94)
              -(2 x 3.65 + 3.71 - (2 x 24.47))
          = -46.21

3600R: $\ln K_p$ = (2 x 22.80 + 32.44 - 2 x 31.96)
              -(2 x 3.70 + 4.25 - (2 x -9.28))
          = -16.09

5400R: $\ln K_p$ = (2 x 24.57 + 34.46 - 2 x 34.69)
              -(2 x 3.93 + 4.48 - (2 x -3.95))
          = -6.02

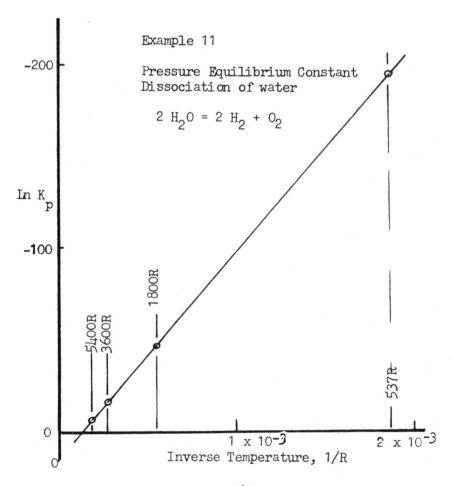

Example 11

Pressure Equilibrium Constant
Dissociation of water

$2 H_2O = 2 H_2 + O_2$

In K_p

-200

-100

0

1×10^{-3} 2×10^{-3}

Inverse Temperature, 1/R

5400R   3600R   1800R   537R

VARIATION OF HEAT OF REACTION WITH TEMPERATURE

The heat of combustion, or heat of reaction, is equal to the difference in the enthalpy of the products and the reactants at a given temperature. The value for the heat of combustion at a different temperature can be found by using the heat of combustion at the given temperature plus the change in the net enthalpy in going to the new temperature.

Consider reactants at constant pressure and changing from temperature 1 to temperature 2. The heat absorbed is the heat capacity multiplied by the temperature rise:

$$dH_1 + C_{p(products)}(T_2-T_1) = dH_2 + C_{p(reactants)}(T_2-T_1)$$

then $(dH_2-dH_1)/(T_2-T_1) = C_{p(products)} - C_{p(reactants)}$

we can also say: $dH = \int_{T_1}^{T_2} \Sigma C_p \; dT$

and since the integral of $C_p \; dT$ is equal to the enthalpy, we can use enthalpy values.

The quantity of heat needed to decompose a substance is the same as the quantity of heat needed for that substances formation, as found by Lavoisier and Laplace in 1780. This is in accord with the conservation of energy, The First Law of Thermodynamics.

Example 12   VARIATION OF HEAT OF COMBUSTION
WITH TEMPERATURE

Determine heat of combustion for hydrogen at 2700R

$$2 \; H_2(g) + O_2(g) = 2 \; H_2O(g) \qquad \text{where g = gas}$$

1.   Find Heat of Formation of substances at 537R

   Water, $H_2O(g)$ H_fo = -57,798 cal/mole
                          (x 1.8= -115,596. BTU/mol)

   Hydrogen, $H_2(g)$ = 0

   Oxygen, $O_2(g)$ = 0

2.   Compute heat of combustion at 537R.

   dH = 2 x(-115,596.)- 2 x (0) - 0
      = -231,192 BTU/formula
         (-231,192/(2 x 2.016= 57,339. BTU/lb)

3.   Find change of enthalpy from 537R to 2700R.
     See BASIC Computer Program 6

   Water, H_2O,   dH = 1147. BTU/lb x 18 = 20,654. BTU/mol
   Hydrogen, H_2 = 7879.   "   x 2. = 15,758.   "
   Oxygen, O_2 = 548.   "   x 32.= 17,536.   "

4.   Find net enthalpy change for formula.

   dH = 2 x 20,654, - 2 x 15,758. - 17,536.
      = - 7,744. BTU/formula

5.   Find heat of combustion at 2700R.

   dH = -231,192 + (-7,744) = -238,936. BTU/formula
        (-238,936/(2 x 2.016= -59,260 BTU/lb)

RELATION BETWEEN TEMPERATURE SLOPE OF EQUILIBRIUM CONSTANT AND HEAT OF REACTION

Consider the van't Hoff equation, which we will differentiate with respect to temperature.

$$- \Delta G = RT \ln K_p \qquad \text{(Eq. 1)}$$

$$- d \Delta G/dT = R \ln K_p + RT \, d(\ln K_p)/dT \qquad \text{(Eq. 2)}$$

Recall the Gibbs-Helmholtz equation of previous derivation

$$(dG/dT)_P = (\Delta G - \Delta H)/T \qquad \text{then multiplying by minus 1}$$

_or_ $\quad - (dG/dT)_P = (\Delta H - \Delta G)/T \qquad \text{(Eq. 3)}$

now substituting,

$$(\Delta H - \Delta G)/T = R \ln K_p + RT \, d(\ln K_p)/dT$$

_or_ $\quad \Delta H - \Delta G = RT \ln K_p + RT^2 \, d(\ln K_p)/dT \quad \text{(Eq. 4)}$

For a special case of importance, which is the standard state (at 1 atmosphere and 25 C) and designated by the superscript zero, thus

$$\Delta G^o = - RT \ln K_p \qquad \text{(Eq. 5)}$$

Now, substituting Equation 5 in Equation 4, the second and third terms drop out, so that

$$d (\ln K_p)/dT = \Delta H^o/RT^2 \qquad \text{(Eq. 6)}$$

This equation can be used to calculate heats of formation from chemical equilibria, and to calculate equilibrium constants at different temperatures. Equation 6 can be integrated by the relation $\int u^n = u^{(n+1)}/(n+1)$.

$$\ln K_p = -(dH/R)(1/T) + C$$

$$\text{or} \quad \ln K_{p2} - \ln K_{p1} = -(dH/R)(1/T_2 - 1/T_1) \quad \text{(Eq. 7)}$$

Example 13. FINDING HEAT OF COMBUSTION FROM
SLOPE OF EQUILIBRIUM CONSTANT

Formation of Ammonia at One Atmosphere

$$0.5 \ N_2 + 1.5 \ H_2 = NH_3$$

Use data from Example 8.

K_p at 2000R = -8.10
at 1000R = -2.10   Find Heat of Formation at 1500R

1. Use Equation 7, above.

$$\ln \ K_{p2} - \ln K_{p1} = - \ dH^o/R \ (1/T_2 - 1/T_1)$$

$$-8.10 - (-2.19) = - \ dH^o/1.986(1/2000 - 1/1000)$$

$$- \ dH^o = -5.91 \ x \ 1.986/(1/2000 - 1/1000)$$

$$dH^o = - \ 23,475. \ BTU/mol$$

Example 14. FINDING SLOPE OF PRESSURE EQUILIBRIUM CONSTANT
FROM HEAT OF FORMATION

Formation of Ammonia at One Atmosphere

Use data from Example 13, and Equation 7, above.

1. Find T_2 with T_1 = 1000R and dH^o = -23,475. BTU/mol

$$1/T_2 - 1/T_1 = (\ln K_{p2} - \ln K_{p1}) \ R/(-dH^o)$$

$$1/T_2 - 1/1000 = ((-8.10 - (-2.19)) x \ 1.986)/(\ 23,475)$$

$$1/T_2 = - \ 5.91 x 1.986/23475) + (1/1000)$$

$$= 5.000 \ x \ 10^{-4}$$

$$T_2 = 2000R$$

CHEMICAL EQUILIBRIA

Chemical equilibria deals most particularly with liquids, for it was with this phase of matter that many of its laws were developed. There is a close similarity of chemical actions in liquids and in gases, though not always apparent.

The true nature of water, that it is composed of hydrogen and oxygen; and of air, that it is composed of oxygen and a larger fraction of inert gas, was first uncovered by Henry Cavendish (1731-1810). Cavendish experimented with the gases evolved by addition of metal chips to acid. His discoveries were challenged for authorship by Antoine Lavoisier (1743-1794), a victim of the guillotine. There were, indeed, hundreds of scientists in this era working to explain questions of science.

Electrolysis, the flow of electricity in fluids, was studied by Michael Farady (1791-1867) who formulated many of its laws. Nevertheless, he failed to perceive the underlying nature of its laws.

Svante Arrhenius (1859-1927), as a student, was faced with the paradox that an electric current could not be made to flow in distilled water, nor in common table salt, yet, when salt was added to the water a current was obtained.

In 1883, at the age of 24, the explanation occured to him. His theory was considered so unlikely that it was rejected by his colleagues and only eventually accepted through support by his friends Wilhelm Ostwald (1853-1932) and Jacobus van't Hoff (1852-1911). Each of the three won Nobel prizes for research in this area.

THE ARRHENIUS THEORY OF ELECTROLYTIC DISSOCIATION

1. An electrolytic molecule spontaneously dissociates in a solvent into two or more particles called ions which bear opposite electrical charges. Positively charged particles are called cations and negatively charged particles are called anions.

2. Ions behave like molecules of solute in determining the physical properties of solution.

3. The dissociation is not complete.

4. An electrical current through an electrolyte consists of the motion of ions under the influence of the electrical field.

5. The mobility of an ion, its velocity in centimeters per second at a potential of one volt per centimeter, is independent of the electrolyte concentration.

SOLUTIONS

Solutions can exist as gas, liquid or solid. Where there are two constituents, the larger quantity is called the solvent and the lesser the solute.When solids or gases are dissolved in a liquid, the latter is called the solvent. A saturated solution is one in which the solute is in equilibrium with the solvent. A larger quantity of a solid solute will precipitate from a saturated solution; a lesser quantity will dissolve.

RAOULT'S LAW

A solute added to a solvent will lower its vapor pressure, which simultaneously lowers the freeze point and raises the boiling point.

There are two classes of solvents: electrolytes, and non-electrolytes.

The change in vapor pressure of non-electrolytes can be predicted by Raoult's Law. Raoult (Francois Marie,1830-1901) found that the lowering of the vapor pressure of a solvent is proportional to the molar concentration of the substance dissolved. One mole per liter of a substance lowers the freezing point by 1.855 deg C. Non-electrolytes, for which the Law applies, include cane sugar and glycerine.

$$P_s = N_B \ p^o$$

where P_s = Vapor pressure of solvent in the solution
N = Mole fraction of a non-volatile solute
p^o = Vapor pressure of the pure solvent.

The change in vapor pressure for electrolytes in solution is greater than for non-electrolytes. This class includes acids, salts and bases: sodium cloride, NaCl, Potassium Nitrate, KNO_3, and Magnesium Sulfate, $MgSO_4$.

Raoult's Law applies to dilute mixtures.

HENRY'S LAW

William Henry (1774-1836) found that the amount of gas dissolved in a liquid is proportional to the pressure of the gas above the liquid surface.

$$P_{solute} = K \ N_{solute}$$

where P_{solute} = Vapor pressure of solute
K = Solubility factor,
N_{solute} = Mole fraction of solute

Note that Henry's Law applies to the solute; Raoult's to the solvent.

APPLICATION OF THE ARRHENIUS THEORY TO FINDING CONCENTRATION OF SOLUTE

The first three statements of the Arrhenius Theory listed above can be used to express mathematically the quantity of a solute in solution.

Consider a binary electrolyte

$$HCl = H + Cl$$

The number of ions produced is 2, or $v = 2$

Consider another electrolyte

$$BaCl_2 = Ba + 2 Cl$$

The number of ions produced, $v = 3$

Let n = the molar concentration

Let α equal the fraction of the molecules that are dissociated

Then, the fraction of the molecules which are not dissociated is $n-1$, and the concentration of the undissociated molecules is $\alpha(n-1)$

The total number of moles of solute will equal

$$n(1 - \alpha) + v\, n\, \alpha$$

Call abnormality, i, the ratio of the number of moles in solute to the number of moles in solution for no dissociation.

Then, $i = (n(1 - \alpha) + vn\alpha)/n$

67

Example 15. LOWERING OF FREEZE POINT OF MIXTURE

Find the freeze point lowering of 0.5 mole of
salt, NaCl, in 1000 grams of water (one liter)
with the salt 77 percent ionized.

PROCEDURE:

1. Find the concentration of ions.

$$c_i = v \times n \times \alpha$$

where c_i = concentration of ions
v = number of ions = 2
n = molar concentration = 0.5 per liter
α = fraction dissociated = 0.77

then, c_i = 2 x 0.5 x 0.77 = 0.77 mole per liter

2. Find total concentration of undissociated molecules, c_u.

$$c_u = n(1 - \alpha)$$

$$= 0.5 (1 - 0.77) = 0.115 \text{ mole per liter}$$

3. Find total concentration of ions and undissociated
molecules, c.

$$c = c_i + c_u$$

$$= 0.77 + 0.115 = 0.885 \text{ mole per liter}$$

4. Find Degree of Abnormality, A.

(With no dissociation, concentration would be
0.5 mole/liter.)

$A = c/n$
$= 0.885/0.5 = 1.77$

5. Find freeze point lowering of solution, dF

(Note that one mole per liter lowers the freeze point
by 1.855 deg C.)

$dF = n \times (-1.855) \times A$
$= 0.5 \times (-1.855) \times 1.77$
$= -1.63 \text{ deg C}$

THE EQUILIBRIUM CONSTANT EXPRESSED AS CONCENTRATION

Consider the general reaction

$$aA + bB + \ldots = gG + hH + \ldots$$

where A and B are the reactants and G and H are the products and a,b,g and h are the moles of concentration, which is express= ed as moles per liter.

Note that at the start of a reaction, the product concentration is zero under the usual conditions.

At equilibrium, the products are being formed at the same rate at which the reactants are being reformed.

If $a = b = g = h$, then the equilibrium constant for concentration, K_c, can be expressed as

$K_c = 1$, where K_c is the ratio of the concentration of the products to the reactants

Where the molar concentrations are not equal (the usual case) the _effective_ concentrations are found to be equal to the actual concentrations to the exponential value set by the equation coefficient.

$$K_c = \frac{(C_G)^g \times (C_H)^h}{(C_A)^a \times (C_B)^b}$$

Note that the two sides of the equation can produce an equation such that K_c is not dimensionless. Its value is usually expressed as moles per liter.

Example 16. FIND EQUILIBRIUM CONSTANT OF CONCENTRATION

Reaction of Hydrogen and Iodine   at 448 deg C

$$H_2 + I_2 = 2 \; HI$$

Concentration of H_2 = 0.0022 moles/liter
          (0.0022 x 2.016 = 0.0044 gm/liter)
      I_2 = 0.0022 moles/liter
          (0.0022 x 235.84 = 0.5584 gm/liter)
      HI = 0.0156 moles/liter
          (0.0156 x 127.928 = 1.9957 gm/liter)

PROCEDURE:

1.   Use equilibrium constant equation for concentration.

$$K_c = (C_G)^g \; x \; (C_H)^h \; / ((C_A)^a \; x \; (C_B)^b)$$

$$= (0.0156)^2 / ((0.0022) x (0.0022))$$
$$= 50.2810 \; moles/liter$$

2.   Note that if the concentrations are doubled, K_c is constant.

3.   Note that for the reverse reaction, K_c is the inverse.

$$2 \; HI = H_2 + I_2$$

$$K_c = 1/50.2810$$
$$= 0.0199$$

Is is seen that the equilibrium concentration constant, K_c, remains constant in the above experiment. Such a relationship was enunciated in 1867 by Guldberg and Waage as The Law of Mass Action.

> The rate (that is, the fraction of substance reacted) of a chemical reaction is proportional to the active masses of the reacting substances present at the time.

The process of reaction to form products can become reversed if the concentration of the products becomes intense. This was first noted by Wenzel in 1777. Claude-Louis Berthellet (1748-1822), as an advisor to Napoleon in Egypt, saw evidence of it in 1799 on the shore of certain lakes in the area. He found sodium carbonate, Na_2CO_3, in evidence, despite the normal conversion to calcium carbonate, $CaCO_3$.

$$Na_2CO_3 + CaCl_2 = 2 NaCl + CaCO_3$$

PRINCIPLE OF LE CHATELIER

Henri-Louis Le Chatelier (1850-1936) studied equilibrium processes as they apply to chemical physics and evolved the principle which bears his name.

> The displacement from equilibrium of a process in a given direction is opposed by a generated force which tends to oppose and nullify that movement.

Consider the reaction which has an unequal number of moles for reactors and products, where n is the value of the reactor A, and Q is the heat evolved by the products:

$$n A + B \rightleftharpoons C + D + Q$$

(1) An increase in temperature displaces the reaction to the left.
(The reaction absorbs heat in that direction.)

(2) In increase in pressure displaces the reaction to the right.
(The right side had less volume which tends to absorb the pressure increase.)

(3) An increase in the concentration of A or B moves the reaction to the right; a decrease in the concentration of A or B moves the reaction to the left.

Example 17. RELATION BETWEEN PRESSURE EQUILIBRIUM
CONSTANT AND DISSOCIATION

Dissociation of Nitrogen Tetroxide

$$N_2O_4 = 2\ NO_2$$

At 25°C and One Atmosphere, N_2O_4 was found to be
18.46 % dissociated.
Find K_p.

PROCEDURE:

1. Let α = fraction of dissociation

    then, $P_{NO_2} = (2\alpha/(1+\alpha))P$

    $P_{N_2O_4} = ((1-\alpha)/(1+\alpha))P$

2. The pressure equilibrium constant is:

    $$K_p = (P_{NO_2})^2/(P_{N_2O_4})$$

    substituting:

    $$K_p = \frac{(2\alpha/(1+\alpha)P)^2}{((1-\alpha)/(1+\alpha))P}$$

    $$= \frac{4\alpha^2 P^2}{(1-\alpha)(1+\alpha)}$$

    $$= \frac{4\alpha^2 P}{1-\alpha^2}$$

3. Now, substitute values of the problem:

    $$K_p = (4(0.1846)^2/(1-(0.1846)^2) \times 1\ \text{atm.}$$

    $$= 0.141\ \text{atm.}$$

EQUILIBRIUM CONSTANT AS MOLE FRACTION

The equilibrium constant can be expressed as a mole fraction by similarly considering the mole fractions of each constituent to the exponential power of the equation coefficient.

$$aA + bB + \ldots = gG + hH + \ldots$$

$$K_X = \frac{(X_G)^g \times (X_H)^h}{(X_A)^a \times (X_B)^b}$$

The mole fraction equilibrium constant is related to the pressure equilibrium constant as follows:

$$K_p = \frac{(X_G P)^g \times (X_H P)^h}{(X_A P)^a \times (X_B P)^b} = \frac{(X_G)^g \times (X_H)^h}{(X_A)^a \times (X_B)^b} P^{(g+h) - (a+b)}$$

$$K_p = K_x P^{dn}$$

note that the value of K_x depends on the pressure if dn is not equal to zero.

RELATION OF GIBBS ENERGY TO TYPE OF EQUILIBRIUM CONSTANT

The general formula for relating Gibbs energy to the equilibrium constant is:

$$dG = -RT \ln K$$

where K can be expressed as either concentration or pressure

however, the values of dG so determined will not be the same if the sum of the volumes of the reactants and the products is not the same, that is, if dn is not equal to zero.

EQUILIBRIUM CONSTANT EXPRESSED AS PRESSURE

At any given temperature, the concentration of a perfect gas is proportional to its partial pressure. A pressure equilibrium constant can be developed in the same manner as for the concentration equilibrium constant, but using values of pressure rather than concentration.

$$aA + bB + \ldots = gG + hH + \ldots$$

$$K_p = \frac{(P_G)^g \times (P_H)^h}{(P_A)^a \times (P_B)^b}$$

RELATION BETWEEN PRESSURE AND VOLUME EQUILIBRIUM CONSTANTS

By the perfect gas law, the concentration is

$$1/v = P/(RT)$$

here, we must use a value of R of consistency with P in atmospheres

$$R = PV/nT$$

$$= (1 \text{ atm} \times 22.414 \text{ liters/mole})/(1 \text{ mole} \times 298 \text{ K})$$
$$= 0.08205 \text{ liter-atm/mole-K}$$

then,

$$K_c = \left(\frac{1}{RT}\right)^{(g+h)-(a+b)} \times \frac{(P_G)^g \times (P_H)^h}{(P_A)^a \times (P_B)^b}$$

for convenience, let $dn = (g + h) - (a + b)$

then, since $(1/x)^n = x^{-n}$

by reduction, $K_c = (RT)^{-dn} K_p$

likewise, $K_p = K_c (RT)^{dn}$

Observe that where the same number of moles is present in the reactants and the products, that is, where $dn = 0$, then $K_p = K_c$.

Example 18.   RELATION OF EQUILIBRIUM VOLUME CONSTANT
TO EQUILIBRIUM PRESSURE CONSTANT

Formation of Ammonia

$$N_2 + 3 H_2 = 2 NH_3$$

At $400°C$, the equilibrium volume constant, $K_c = 0.507$

PROCEDURE:

1. Find value of the net change in molar volume, dn

    $$dn = 2 - (1 + 3) = -2$$

2. Convert to equilibrium pressure constant, K_p

    $$K_p = (R\ T)^{dn}\ K_c$$

    where R = 0.08205 liter-atm/mol-K
    T = deg K = 273 + deg C

    $$K_p = (0.08205\ x\ (400+273)^{-2}(0.507)$$

    $$= 1.66\ x\ 10^{-4}$$

ln K_p = -8.70 which agrees with calculations of Example 10.

Example 19. TOTAL PRESSURE REQUIRED FOR FORMATION OF
5 PERCENT OF AMMONIA AT 400 C.

$$N_2 + 3\ H_2 = 2\ NH_3$$

By Example 18, K_p = 1.66 x 10^{-4} at 400C

PROCEDURE:

1. Setup partial pressure of each constituent in terms of
final total pressure.

For 1 volume of nitrogen, and 3 volumes of hydrogen,

$$P_{NH_3} = 0.05\ P$$

$$P_{N_2} = (1/(1+3))\ x\ (1-0.05)\ P$$

$$= 0.2375\ P$$

$$P_{H_2} = (3/(1+3))\ x\ (1-0.05)\ P$$

$$= 0.7125\ P$$

2. Substitute values in expression for K_p.

$$K_p = (P_{NH_3})^2/((P_{N_2})(P_{H_2})^3$$

$$1.66\ x\ 10^{-4} = (0.05\ P)^2/((0.2375\ P)(0.7125\ P)^3$$

$$1.66\ x\ 10^{-4} = 0.02912/P^2$$

$$P = (175)^{1/2} = 13.2\ atm.$$

DISSOCIATION OF WATER, THE pH FACTOR

Water is the most commonly used solvent in chemical processing, and has been called the Universal Solvent.

Water is normally partly dissociated.

$$H_2O = H^+ + OH^-$$

$$K = \frac{H^+ \times (OH)^-}{(H_2)}$$

The product $H^+ \times (OH)^- = K_w$

where K_w is called the <u>ion product</u>

The value of K_w for water at 25 C is found to be 1×10^{-14} from electrical conductivity measurements. In pure water, the concentration of H and OH ions is the same, so that the concentration of each is the same, namely 1×10^{-7}.

A neutral solution of water, therefore, has a concentration of 10^{-7} moles per liter.

The acidity or basicity of an aqeous (water) solution can be expressed in terms of the pH.

$$pH = \log (1/H)$$

where H = the Hydrogen Ion concentration, moles/liter

Accordingly, the pH of a neutral solution is

$$pH = \log (1/(1 \times 10^{-7})) = 7.0$$

The concept of pH was introduced in 1909 by Sorensen. It can be seen that acid solutions have a value from 0 to 7; alkalines from 7 to 14.

EQUILIBRIA IN SOLUTIONS OF ELECTROLYTES

The Arrhenius Theory of Electrolytic Dissociation assumes that when electrolytes are dissolved in solvents like water there is partial dissociation. The undissociated molecules are in equilibrium with the dissociated ones, which are called ions.

For the dissociation of compound AB, it may be written

$$AB = \overset{+}{A} + \overset{-}{B}$$ showing that one ion is negative; the other is positive

The ordinary laws of equilibrium are found to apply and for the above example, we can write

$$K = \frac{\overset{+}{A} \times \overset{-}{B}}{(AB)}$$

where K = the dissociation constant
and the values for A, B and AB are the concentration in moles per liter

OSTWALD'S DILUTION LAW

Ostwald found that the dissociation constant, K, is independent of concentration for weak electrolytes (those that do not completely dissociate. Weak electrolytes include acetic acid and ammonium hydroxide.

Strong electrolytes, for which the Ostwald dilution law does not apply, include Potassium Chloride, KCl, Barium Nitrate, $Ba(NO_3)_2$, and Ferricyanide, $K_4 Fe(CN)_4$.

REACTION RATES

TYPES OF REACTIONS

Most reactions are quite complex and consist of a series of steps, some of which are fast and some of which are slow. For example consider the decomposition of ethyl bromide:

$$C_2H_5Br = C_2H_4 + HBr$$

The reaction actually proceeds in this fashion:

$$C_2H_5Br = C_2H_5 + Br$$

$$Br + C_2H_5Br = HBr + C_2H_4Br$$

$$C_2H_4Br = C_2H_4 + Br$$

A reaction whose rate is proportional to the concentration of one reacting substance is called a First Order Reaction. Similarly, if proportional to the concentration of two substances, a Second Order Reaction; if three, a Third Order Reaction. There are, also, Zero Order Reactions, in which the reaction is not affected by concentration. Some complex reactions cannot be so simply categorized.

THE ARRHENIUS EQUATION

Arrhenius found that most reaction rates could be expressed by an exponential relationship, which plots as a straight line on semi-logarithmic paper.

$$K = A\ e^{-(E/RT)}$$

where K = Rate constant
A = Constant for the reaction
e = Base of natural logarithms, 2.718+
R = Gas constant, 8.314 J/mol-deg K
T = Temperature, deg K
E = Activation energy, found by experiment,

Example: $A = 1.0 \times 10^{14}$
$E = 80,000.$
$T = 300\ K$

$$K = (1.0 \times 10^{14})\ e^{-(80000/(8.314 \times 300))}$$
$$= 4.046 \times 10^{12}$$

CATALYSTS

Catalysts are substances which participate in a chemical reaction but are not consumed by it. The catalyst may, however, be fouled by the reaction and need to be cleaned or replaced.

There are classes of catalysts as surface or homogeneous, the distinction being whether area or volume is the controlling factor.

The same reactants may yield one set of products with one catalyst and another set with another catalyst.

Carbon monoxide, CO, reacts with hydrogen, H_2, in the presence of ZnO and Cr_2O_2 to produce methyl alcohol:

$$CO + 2 H_2 \xrightarrow{ZnO, CrO_2} CH_2OH$$

In the presence of Nickel, however it produces methane:

$$CO + 3 H_2 \xrightarrow{Ni} CH_4 + H_2O$$

The autocatalyst, or self-catalyst, is a special type of reaction. For example, the reaction of permanganate ion with oxalic acid:

$$\underset{\text{violet}}{2 \ MnO_4^-} + 5 \ H_2C_2O_4 + 6 \ H_3O^+ = \underset{\text{colorless}}{2 \ Mn^{2+}} + 10 \ CO_2 + 14 \ H_2O$$

The product Mn^{2+} catalyzes the reaction. Addition of a trace of this ion enables the reaction to start.

There are biological catalysts known as enzymes. For instance, Pepsin in the gastric juice, and Ptyalin in the saliva. Ptyalin accelerates the conversion of starch to sugar.

There are numerous examples of catalysts:

NO is used in the preparation of nitric acid.

Platinum aids the reaction of SO_2 with O_2 to form SO_3

Crude oil is catalyzed into lighter ends by Al_2O_3

Nickel catalyzed vegetable oil and hydrogen to form shortenings and oleomargarine.

Platinum gause catalyzes automobile exhaust products into more desirable compounds

Iron and Al_2O_3 are used in making ammonia in the Haber Process.

The Haber Process for the production of ammonia requires the reactants, hydrogen and nitrogen, to be heated to 400-500C and compressed to 100 to 1000 atmospheres over an iron catalyst.

A patented new process for ammonia formation works at room temperature and pressure by use of titanium and zeolites as catalysts.

Zeolites are a class of silicate minerals with a cage-like structure. These cages contain ions which can trap smaller molecules. The trapped molecules can then be subjected to chemical reactions by addition of other chemicals.

The calcium zeolite. when washed by titanium chloride retains some free titanium. The modified zeolite can then form free hydrogen and oxygen under the influence of visible light. The next step consists of bubbling nitrogen through the water which produces the ammonia under the action of the catalyst. Yields are not commercial, and titanium is consumed in the process.

Mobil has developed a process using zeolite catalysts to transform methanol (methyl alcohol) into liquid fuels.

Exxon has discovered a process that uses potassium carbonate to help turn coal into gas. Exxon is planning a $500 million pilot plant in Rotterdam by 1985. Mobil is considering a plant in New Zealand based on the technique.

INHIBITORS

There are certain substances which act to reduce the speed of a chemical reaction.

This effect has been well researched in relation to "knock" in spark ignition engines, which is related to detonation and the speed of combustion.

For a given number of carbon atoms in the molecule, anti-knock value is poorest for straight-chain (paraffin) molecules, and improves sa the carbon atom arrangement becomes more complex and as the carbon-hydrogen ratio becomes larger.

Anti-knock additives to gasoline were investigated in 1920 by T. Midgeley, Jr. The relative effectiveness of various compounds is listed below:

EFFECTIVENESS OF ANTI-KNOCK COMPOUNDS

COMPOUND	FORMULA	RELATIVE MOL. EFFECTIV.
Tetraethyl Lead	$(C_2H_5)_4Pb$	114.
Diphenyl Diethyl Lead	$(C_6H_5)_2(C_2H_5)_2Pb$	110.
Nickel Carbonyl	$Ni(CO)_4$	35.
Tetraphenyl Lead	$(C_6H_5)_4Pb$	69.5
Aniline	$C_6H_5NH_2$	1.
* Isopropyl Nitrate	$C_3H_7NO_2$	11.5

* Pro-knock compound

Hydrogen peroxide decomposes into water and oxygen at such a rate that storage is difficult. Decomposition can be slowed by the addition of phosphates. It is thought that the action of the inhibitor is to destroy such catalysts as may be present.

REFERENCES

1   Organic Chemistry, Keith M. Seymour, Prentice-Hall, Inc., Englewood Cliffs, New Jersey, 1961

2   Applied Hydrocarbon Thermodynamics, Wayne C. Edmister, Gulf Publishing Co., Houston, Texas, 1961

3   Thermochemische Untersuchungen (Inquiry into Chemistry), Julius Thomsen, Leipzig, 1886

4   Thermochemistry, J. Thomsen, London 1908 (Translation of 3)

5   On the Calculation of Thermo-Chemical Constants, H. Stanley Redgrove, London, Edward Arnold, 1909

6   The Strengths of Chemical Bonds, T. L. Cottrell, Butterworths Scientific Publications, London, Second Edition, 1958

7   Selected Values of Properties of Hydrocarbons, F. D. Rossini et al, U.S. Dept. of Commerce, NBS C461, Washington, DC, 1947

8   JANAF Thermochemical Tables, D.R. Stull and H. Prophet, US Dept of Commerce, NSRDS-NBS 37, US Govt Printing Office, 1971

9   Gas Power Dynamics, A.D. Lewis, D van Nostrand, New Jersey, Princeton, 1962

10 Physical Chemistry, F. Daniels and R. Alberty, John Wiley & Sons, New York, 1966

11 Chemistry, J. V. Quagliano, Prentice-Hall, New Jersey, 1958

12 Principles of Chemistry, J. Hildebrand, The Macmillan Co., New York, 1947

13 General Chemistry, T.P. McCutcheon, H. Seltz, and J. Warner, D. Van Nostrand Co., New York 1939

14 Thermodynamics, Virgil Faires, The MacMillan Co., NY, 1962

15 Outlines of Physical Chemistry, Farrington Daniels (Fred. H. Getman), John Wiley and Sons, New York, 1945

16 Chemical Principles and Processes, M.J. Sienko and R. Plane, McGraw-Hill Book Co., New York, 1974

17 Chemical Process Principles, O.A. Hougen, K. M. Watson, and R. Ragatz, John Wiley & Sons, New York, 1954

APPENDIX I

BASIC COMPUTER PROGRAMS

1. REAL GAS, for compressibility specific weight

2. 3RD ORDER, solves by determinants three sets of equations

3. SPECIFIC HEAT, CP, finds specific heat at constant pressure

4. HEAT OF COMBUST, finds heat of combustion for hydrocarbons

5. HEAT OF FORM, finds heat of formation for hydrocarbons

6. ENTHALPY, uses values of C_p to find Enthalpy

7. ENTROPY, uses values of C_p to find Entropy

8. ISOTH EXP V, finds work of isothermal expansion for known change of volume

9. ISOTH EXP P, finds work of isothermal expansion for known change of pressure

10. ADIAB EXP T, finds final temperature for adiabatic expansion

11. ADIAB EXP V, finds final volume for adiabatic expansion

12. ADIAB EXP P, finds final pressure for adiabatic expansion

13. PRESS EQUIL GIBBS, finds equilibrium pressure constant for a system having two reactants and one product

14. PLOT, produces a typewriter output terminal or CRT display plot of two sets of data on X,Y axes.

15. PLOT KP, produces a display of two sets of data for temperature and pressure equilibrium constant, plotting ln K_p vs. 1/T.

```
10    REM   "REAL GAS"

20    DIM C(4)

30    DIM X(2)

40    PRINT "INPUT CRIT. TEMP, R"

50    INPUT TC

60    PRINT "INPUT CRIT. PRESS., PSIA"

70    INPUT PC

80    PRINT "INPUT NORMAL BOIL, F"

90    INPUT TNB

100   PRINT "INPUT PRESS, PSIA"

110   INPUT P

120   PRINT "INPUT TEMP, F"

130   INPUT T

140   PRINT "INPUT MOL. WT."

150   INPUT MOLWT

160   LET SPWT = (P * 144)/((1545/MOLWT) * (T + 460))

170   LET TR = (T + 460)/TC

180   LET PR = P/PC

190   PRINT "PR = ", PR,"TR = ", TR,

200   PRINT "INPUT COEFF. FOR SIMPLE FLUID, 4 X"

210   FOR N = 1 TO 4

220   INPUT C(N)

230   NEXT N

240   LET ZL = 0.1742 * TR ** 0.8546

250   LET DZV = 1+ C(1) * PR + C(2) * PR ** 2 + C(3) * PR **3
                 + C(4) * PR ** 4
```

("REAL GAS, continued)

260 LET ZO = ZL + DZV

270 PRINT "INPUT CORRECTION COEFF. FOR SIMPLE FLUID, 2X"

280 FOR K = 1 TO 2

290 INPUT X(K)

300 NEXT K

310 LET Z1 = X(1) + (X(2) * TR)

320 LET W = (3/7) * (((IN(PC/14.69))/2.303)/((TC/TNB)-1))-1

330 LET Z = ZO + (W * Z1)

340 LET RGSPWT = SPWT/Z

350 LET FUG = RGSPWT * (1545/MOLWT) * (T + 460)/144

360 PRINT "IDEAL GAS SPECIFIC WT = ", SPWT, "LB/CUFT"

370 PRINT "REAL GAS SPECIFIC WT = ", RGSPWT, "LB/CUFT"

380 PRINT "FUGACITY PRESSURE = ", FUG, "PSIA"

390 GOTO 100

Input	Output
Critical Temp, R = 1161	PR = 0.8138
	TR = 0.9767
Critical Press, psi, = 3195	
	Ideal Gas, sp. wt = 3.85 lb/cuft
Normal Boiling, F ~ = 212	
	Real Gas, sp. wt. = 7.99 lb/cuft
Pressure, PSIA = 2600	
	Fugacity press = 5404. psia
Temp. F = 674	
Mol. Wt. = 18.0	

Coeff. for simple fluid
   C 1 = 0.7653
   C 2 = -4.0547
   C 3 = 3.1614
   C 4 = -0.7234
Correct. for simple fluid
   X 1 = 0.0137
   X 2 = 0.0041

```
1Ø    REM " 3RD ORDER"

2Ø    DIM S(3)

3Ø    DIM  A(3)

4Ø    DIM B(3)

5Ø    DIM C(3)

6Ø    DIM T(3)

7Ø    DIM Q(3)

8Ø    PRINT "INPUT S; INPUT T, 3X"

9Ø    FOR N = 1 TO 3

1ØØ   INPUT S(N)

11Ø   INPUT T(N)

12Ø   NEXT N

13Ø   LET A(1) = T(1)

14Ø   LET B(1) = T(1)** 2

15Ø   LET C(1) = T(1) ** 3

16Ø   LET A(2) = T(2)

17Ø   LET B(2) = T(2) ** 2

18Ø   LET C(2) = T(2) ** 3

19Ø   LET A(3) = T(3)

2ØØ   LET B(3) = T(3) ** 2

21Ø   LET C(3) = T(3) ** 3

22Ø   LET DO = (A(1) * B(2) * C(3)) + (A(2) * B(3) * C(1)) + (A(3)
          * C(2) * B(1)) - (C(1) * B(2) * A(3)) - (B(1) *
          A(2) * C(3)) - (A(1) * B(3) * C(2))
```

230   LET DA = (S(1) $*$ B(2) $*$ C(3)) + (S(2) $*$ B(3) $*$ C(1)) + (S(3) $*$ C(2) $*$ B(1)) - (C(1) $*$ B(2) $*$ S(3)) - (B(1) $*$ S(2) $*$ C(3)) - (S(1) $*$ B(3) $*$ C(2))

240   LET AD = DA/DO

250   LET DB = (A(1) $*$ S(2) $*$ C(3)) + (A(2) $*$ S(3) $*$ C(1)) + (A(3) $*$ C(2) $*$ S(1)) - (C(1) $*$ S(2) $*$ A(3)) - (S(1) $*$ A(2) $*$ C(3)) - (A(1) $*$ S(3) $*$ C(2))

260   LET BD = DB/DO

270   LET DC = (A(1) $*$ B(2) $*$ S(3)) + (A(2) $*$ B(3) $*$ S(1)) + (A(3) $*$ S(2) $*$ B(1)) - (S(1) $*$ B(2) $*$ A(3)) - (B(1) $*$ A(2) $*$ S(3)) - (A(1) $*$ B(3) $*$ S(2))

280   LET CD = DC/DO

290   FOR N = 1 TO 3

300   LET Q(N) = AD $*$ T(N) + BD $*$ T(N) $**$ 2 + CD $*$ T(N) $**$ 3

310   NEXT N

320   PRINT "C1=", AD, "C2=", BD, "C3=",CD

330   PRINT "S1=", Q(1), "INPUT =", S(1), "S2=", Q(2), "INPUT =", S(2), "S3=", Q(3), "INPUT =", S(3)

OXYGEN, O2, GAS

INPUT	OUTPUT
S(1) = 0.2191 BTU/lb-R	C1 = 6.1721 $*$ 10 $**$ -4
T(1) = 537 R	C2 = -4.3916 $*$ 10 $**$ -7
S(2) = 0.2524	C3 = 9.2347 $*$ 10 $**$ -11
T(2) = 1460	Q(1) = 0.2191
S(3) = 0.2725	Q(2) = 0.2524
T(3) = 2660	Q(3) = 0.2725

(3RD ORDER, continued)

```
1Ø    REM "SPECIFIC HEAT, CP"

2Ø    DIM C(5)

3Ø    PRINT "INPUT NAME, 2Ø SPACES"

4Ø    INPUT A$

5Ø    PRINT "INPUT CP, 5 VALUES"

6Ø    FOR N = 1 TO 5

7Ø    INPUT C(N)

8Ø    NEXT N

9Ø    PRINT A$

1ØØ   PRINT "CP 1 = ", C(1), "CP 2= ", C(2), "CP 3 = ", C(3),
         "CP 4 = ", C(4), "CP 5 = ", C(5)

11Ø   PRINT "INPUT TEMP"

12Ø   INPUT TEMP

13Ø   LET CP = C(1) + (C(2) * TEMP) + (C(3) * TEMP **2)
         +(C(4) * TEMP ** 3) + (C(5) * TEMP ** 4)

14Ø   PRINT "TEMP, R =", TEMP,"CP, BTU/LB-R = ", CP

15Ø   GOTO 11Ø
```

OXYGEN, O2, GAS

```
CP 1=    Ø.218
CP 2=    -1.167 * 1Ø ** -5
CP 3=    2.743 * 1Ø ** -8
CP 4=    -4.439 * 1Ø ** -12
CP 5=    Ø
```

TEMP, R=		CP, BTU/LB-R =	
	537		Ø.2189
	96Ø		Ø.2282
	146Ø		Ø.2456
	266Ø		Ø.2885

```
10    REM "HEAT OF COMBUST"

20    PRINT "INPUT NO. OF CARBON ATOMS"

30    INPUT N

40    PRINT "INPUT NO. OF HYDROGEN ATOMS JOINED TO NON-ADJACENT
           CARBON ATOMS"

50    INPUT P

60    PRINT "INPUT NO. OF DOUBLE CARBON LINKS"

70    INPUT A

80    PRINT "INPUT NO. OF TREBLE CARBON LINKS"

90    INPUT B

100   LET HC = N * 210.8 - ( N + P - A - 1 ) * 52.5 - A * 89.0
           - B * 113.0

110   PRINT "HEAT OF COMBUSTION = ", HC, "KILO-CALORIES/GM-MOLE
                                AT 18 C"
120   GOTO 20
```

ETHANE, C_2H_6

Input	Output

Input

    N = 2

    P = 0

    A = 0

    B = 0

Output

    HC = 369.1 KCAL/GM-MOLE
            AT 18 C

```
1Ø    REM    "HEAT OF FORM"

2Ø    PRINT "INPUT NO. OF CARBON ATOMS"

3Ø    INPUT N

4Ø    PRINT  "INPUT NO. OF HYDROGEN ATOMS JOINED TO NON-ADJACENT
                  CARBON ATOMS"

5Ø    INPUT P

6Ø    PRINT "INPUT NO. OF DOUBLE CARBON LINKS"

7Ø    INPUT A

8Ø    PRINT "INPUT NO. OF TREBLE CARBON LINKS"

9Ø    INPUT B

1ØØ   LET HF = N * 21.2 - (N + P - A - 1 ) * 15.Ø - A * 46.Ø
                  - B * 89.5

11Ø   PRINT "HEAT OF FORMATION = ", HF, "KILOCAL/GM-MOLE AT 18 C"

12Ø   GOTO 2Ø
```

ETHANE, C_2H_6

Input   Output

    N = 2   HF = 27.4 KCAL/GM-MOLE AT 18 C

    P = 0

    A = 0

    B = 0

```
10    REM "ENTHALPY"

20    DIM C(5)

30    PRINT "INPUT NAME, 20 SPACES"

40    INPUT A$

50    PRINT "INPUT CP, 5 VALUES"

60    FOR N = 1 TO 5

70    INPUT C(N)

80    NEXT N

90    PRINT A$

100   PRINT "CP 1 =", C(1), "CP 2 = ", C(2), "CP 3 = ", C(3),
         "CP 4 = ", C(4), "CP 5 = ", C(5)

110   PRINT "INPUT TEMP"

120   INPUT T

130   LET ENT = C(1) * T + (1/2) * C(2) * T ** 2 + (1/3) * C(3)
         * T ** 3 + (1/4) * C(4) * T ** 4 + (1/5) * C(5) * T ** 5

140   PRINT "TEMP, R = ", T, "ENTHALPY, BTU/LB = ", ENT

150   GOTO 110
```

OXYGEN, O2, GAS

```
CP 1 =    0.218
CP 2 =    -1.167 * 10 ** -5
CP 3 =    2.743 * 10 ** -8
CP 4 =    -4.439 * 10 ** -12
```

TEMP, R =		ENTHALPY, BTU/LB =	
	537		116.71
	960		211.05
	1460		329.26
	2660		655.12

```
1Ø    REM "ENTROPY"

2Ø    DIM C(5)

3Ø    PRINT "INPUT NAME, 2Ø SPACES"

4Ø    INPUT A$

5Ø    PRINT "INPUT CP, 5 VALUES"

6Ø    FOR N = 1 TO 5

7Ø    INPUT C(N)

8Ø    NEXT N

9Ø    PRINT A$

1ØØ   PRINT "CP 1 = ", C(1), "CP 2 = ", C(2), "CP 3 = ", C(3),
         "CP 4 = ", C(4), "CP 5 = ", C(5)

11Ø   PRINT "INPUT TEMP"

12Ø   INPUT T

13Ø   LET ENT = C(1) * IN T + C(2) * T + (1/2) * C(3) * T ** 2
         + (1/3) * C(4) * T ** 3 + (1/4) * C(5) * T ** 4

14Ø   PRINT "TEMP, R = ", T, "ENTROFY, BTU/IB - R = ", ENT

15Ø   GOTO 11Ø
```

OXYGEN, O2, GAS

CP 1 = Ø.218
CP 2 = -1.167 * 1Ø ** -5
CP 3 = 2.743 * 1Ø ** -8
CP 4 = -4.439 * 1Ø ** -12
CP 5 = Ø

TEMP, R =	ENTROPY, BTU/IB - R =
537	1.368
96Ø	1.497
146Ø	1.596
266Ø	1.757

```
1Ø     REM "ISOTH EXP V"

2Ø     PRINT "INPUT MOLWT"

3Ø     INPUT MOLWT

4Ø     PRINT "INPUT V1, CUFT"

5Ø     INPUT V1

6Ø     PRINT "INPUT V2, CUFT"

7Ø     INPUT V2

8Ø     LET R = V2/V1

9Ø     PRINT "INPUT TEMP, DEG R"

1ØØ    INPUT T

11Ø    LET W = 1.986 * MOLWT * T * LN R

12Ø    PRINT "WORK = ", W, "BTU/LB"

13Ø    GOTO 4Ø
```

AIR

Input

    MOLWT = 28.96

    V1 = 1

    V2 = 10

    T = 491.6

Output

    Work = 67127. BTU/lb

```
10    REM "ISOTH EXP P"

20    PRINT "INPUT MOLWT"

30    INPUT MOLWT

40    PRINT "INPUT P1, PSIA"

50    INPUT P1

60    PRINT "INPUT P2, PSIA"

70    INPUT P2

80    LET R = P1/P2

90    PRINT "INPUT T, DEG R"

100   INPUT T

110   LET W = 1.986 * MOLWT * T * LN R

120   PRINT "WORK = ", W, "BTU/LB"

130   GOTO 40
```

AIR

Input	Output
MOLWT = 28.96	Work = 65103 BTU/lb
P1 = 14.69	
P2 = 1.469	
T = 491.6	

```
1Ø    REM "ADIAB EXP T"

2Ø    PRINT "INPUT MOL WT"

3Ø    INPUT MOLWT

4Ø    PRINT "INPUT SPECIFIC HEAT, CP, BTU/LB-R"

5Ø    INPUR CP

6Ø    PRINT "INPUT INITIAL TEMP, R"

7Ø    INPUT T1

8Ø    PRINT "INPUT EXPANSION RATIO, V2/V1"

9Ø    INPUT R

1ØØ   LET CV = CP - (1.986  /MOLWT)

11Ø   LET X = - (1.986 * LN R)/(CV * MOLWT)

12Ø   LET NUX = EXP X

13Ø   LET T2 = T1 * NUX

14Ø   PRINT "FINAL TEMP = ", T2, "DEG R "

15Ø   GOTO 6Ø
```

AIR

Input

    MOLWT = 28.96

    CP = 0.240

    T1 = 520

    Exp. Ratio = 5.

Output

    Final Temp., T2 = 273. R

```
1Ø    REM "ADIAB EXP V"

2Ø    PRINT "INPUT MOL WT"

3Ø    INPUT MOLWT

4Ø    PRINT "INPUT SPECIFIC HEAT, CP, BTU/LB-R"

5Ø    INPUT CP

6Ø    PRINT INPUT INITIAL VOLUME, V1, CUFT"

7Ø    INPUT V1

8Ø    PRINT "INPUT TEMPERATURE RATIO, T2/T1"

9Ø    INPUT R

1ØØ   LET CV = CP - (1.986/MOLWT)

11Ø   LET X = - (CP * MOLWT * IN R)/1.986

12Ø   LET NUX = EXP X

13Ø   LET V2 = NUX * V1

14Ø   PRINT "FINAL VOL, V2 =", V2, "CUFT"

15Ø   GOTO 6Ø
```

AIR

Input

    MOLWT = 28.96

    CP = 0.240

    V1 = 1

    T2/T1 = 0.525

Output

    V2 = 5.00

```
1Ø     REM "ADIAB EXP P"

2Ø     PRINT "INPUT P1, PSIA"

3Ø     INPUT P1

4Ø     PRINT "INPUT V1, CUFT"

5Ø     INPUT V1

6Ø     PRINT "INPUT V2, CUFT"

7Ø     INPUT V2

8Ø     PRINT "INPUT K (= CP/CV)"

9Ø     INPUT K

1ØØ    LET P2 = P1 * (V1/V2) ** K

11Ø    PRINT "FINAL PRESS, P2, = ", P2, "PSIA"

12Ø    GOTO 2Ø
```

AIR

INPUT   OUTPUT

    P1 = 14.69   P2 = 368.9

    V1 = 10.

    V2 = 1.

    K = 1.40

```
1Ø     REM "PRESS EQUIL GIBBS"

2Ø     DIM A$(3,10)

3Ø     DIM B(3)

4Ø     DIM C(3)

5Ø     DIM D(3)

6Ø     DIM E(3)

7Ø     DIM F(3)

8Ø     DIM G(3)

9Ø     DIM H(3)

1ØØ    DIM J(3,5)

11Ø    DIM K(2,3)

12Ø    DIM L(2,3)

13Ø    DIM Q(3)

2ØØ    PRINT "INPUT NAME PROD, 5 CP, REACT 1, 5  CP, REACT 2, 5 CP"

21Ø    FOR M = 1 TO 3

22Ø    INPUT A$(M,1 TO 10)

23Ø    FOR N = 1 TO 5

24Ø    INPUT J(M,N)

245    PRINT A$(M, 1 TO 10)

25Ø    NEXT N

26Ø    NEXT M

27Ø    PRINT "INPUT MOLES, MOLWT, HEAT FORM, ENTROPY AT 537R"

28Ø    FOR M = 1 TO 3

29Ø    INPUT B(M)
```

Program 13, continued

```
300   INPUT  Q(M)

310   INPUT C(M)

320   INPUT D(M)

330   PRINT "1"

340   NEXT M

430   PRINT "INPUT TEMP, R"

440   INPUT T

450   LET R = 537

460   LET S = 1

470   GOSUB 1000

480   GOSUB 2000

490   LET R = T

500   LET S = 2

510   GOSUB 1000

520   GOSUB 2000

530   FOR N= 1 TO 3

540   LET E(N) = (K(2,N) - K(1,N)) * Q(N)

550   LET F(N) = (L(2,N) - L(1,N)) * Q(N)

560   LET G(N) = - D(N) - F(N) + E(N)/T

570   NEXT N

580   LET DGT = G(1) * B(1) - G(2) * B(2) - G(3) * B(3)
            + (C(1) - C(2) - C(3))/T

590   PRINT "DH=", E(1), E(2), E(3)

600   PRINT "DS=", F(1), F(2), F(3)

610   PRINT "GIBBS=", G(1), G(2), G(3)
```

Program 13, continued

```
62Ø   PRINT "DGT=", DGT

63Ø   LET IKP = - DGT/1.986

64Ø   LET KP = EXP IKP

65Ø   PRINT "IN KP = ", IKP, "KP =", KP

66Ø   GOTO 43Ø

1ØØØ     REM SUB ENTHALPY

1Ø1Ø     FOR N = 1 TO 3

1Ø2Ø     LET K(S,N) = J(N,1) * R + (1/2) * J(N,2) * R ** 2
             +(1/3) * J(N,3) * R ** 3 + (1/4) * J(N,4) *R ** 4
             +(1/5) * J(N,5) * R ** 5

1Ø3Ø     NEXT N

1Ø4Ø     RETURN


2ØØØ     REM SUB ENTROPY

2Ø1Ø     FOR N = 1 TO 3

2Ø2Ø     LET L(S,N) = J(N,1) * IN R + J(N,2) * R
             + (1/2) * J(N,3) * R ** 2 + (1/3) * J(N,4) * R ** 3
             + (1/4) * J(N,5) * R ** 4

2Ø3Ø     NEXT N

2Ø4Ø     RETURN
```

Program 13, continued

$$0.5 \ N_2 + 1.5 \ H_2 = NH_3$$

INPUT

    PRODUCT = AMMONIA

    CP 1 = Ø.412

    CP 2 = 1.8126 * 1Ø **-4

    CP 3 = 2.6235 * 1Ø **-8

    CP 4 = -7.8169 * 1Ø **-12

    CP 5 = Ø

    REACTOR 1 = NITROGEN

    CP 1 = Ø.248

    CP 2 = -1.Ø65 * 1Ø **-5

    CP 3 = 2.154 * 1Ø **-8

    CP 4 = -3.365 * 1Ø **-12

    CP 5 = Ø

    REACTOR 2 = HYDROGEN

    CP 1 = 2.2

    CP 2 = 3.4429 * 1Ø **-3

    CP 3 = -2.522Ø * 1Ø **-6

    CP 4 = 7.12Ø2 * 1Ø **-1Ø

    CP 5 = -6.5725 * 1Ø **-14

    MOLES, MOLWT, HEAT FORM, ENTROPY

    1
    17.02
    -19746
    46.01   PRINT "1"

Program 13, continued

        Ø.5
        28.Ø1
        Ø
        45.Ø1   PRINT "1"
        1.5
        2.Ø16
        Ø
        31.21   PRINT "1"

        INPUT TEMP =   537

OUTPUT

        DH = Ø, Ø, Ø

        DS = Ø, Ø, Ø

        GIBBS = -46.Ø1, -45.Ø1, -31.21

        DG/T = -13/461

        IN KP = 6.778

        KP = 878.

INPUT TEMP = 1ØØØ

OUTPUT

        DH = 4439, 3258, 3394

        DS = 5.9Ø9, 4.369, 4.536

        GIBBS = -47.48Ø, -46.12Ø, -32.352

        DG/T = 4.362

        IN KP = -2.196

        KP = Ø.1112

INPUT TEMP = 2ØØØ

        DH = 16622, 1Ø811, 1Ø939

        DS = 14.215, 9.571, 9.777

        IN KP = -8.096

        KP = 3.Ø49 * 1Ø **-4

104

```
10    REM "PLOT"

20    DIM A(2,10)

30    DIM B(2,10)

40    DIM C$(5,12)

50    DIM D(2,5)

60    DIM X(2,10)

70    DIM Y(2,10)

80    PRINT "INPUT PLOT NAME"

90    INPUT C$(3,1 TO 12)

100     PRINT "INPUT XMAX"

110     INPUT XMAX

120      PRINT "INPUT YMAX"

130     INPUT YMAX

140     PRINT "INPUT X NAME"

150     INPUT C$(1,1 TO 12)

160     PRINT "INPUT Y NAME"

170     INPUT C$(2,1 TO 12)

180     PRINT "INPUT * DATA"

190     FOR N= 1 TO 10

200     INPUT A(1,N)

210     INPUT B(1,N)
215     PRINT A(1,N); B(1,N)
220     NEXT N

230     PRINT "INPUT 0 DATA"

240     FOR N= 1 TO 10

250     INPUT A(2,N)
```

```
260    INPUT B(2,N)
265    PRINT A(2,N); B(2,N)
270    NEXT N

280    PRINT "INPUT * DATA NAME"

290    INPUT C$(4,1 TO 12)

300    PRINT "INPUT Ø DATA NAME"

310    INPUT C$(5,1 TO 12)

320    LET D(1,1) = Ø

330    LET D(2,1) = Ø

340    FOR N = 2 TO 5

350    LET D(1,N) = D(1,(N-1)) + (1/5) * XMAX

350    LET D(2,N) = D(2,(N-1)) + (1/5) * YMAX

360    NEXT N

370    CLS

380    PRINT AT Ø,1Ø; C$(3,1 TO 12)

390    PRINT AT 19,Ø; "I"; TAB 6;"I"; TAB 12;"I"; TAB 18;"I";
              TAB 24;"I"; TAB 30; "I";

400    PRINT AT 2Ø,Ø; D(1,1); TAB 6; D(1,2); TAB 12; D(1,3);
              TAB 18; D(1,4); TAB 24; D(1,5);TAB 30; XMAX

410    PRINT AT 21,12; C$(1,1 TO 12)

420    PRINT AT 4,5;"-"; AT 7,5;"-"; AT 1Ø,5; "-"; AT 13,5;
              "-"; AT 16,5; "-";AT 19,5;"-";

430    PRINT AT 4,Ø; YMAX  ; AT 7,Ø; D(2,5); AT 1Ø,Ø; D(2,4);
              AT 13,Ø; D(2,3);AT 16, Ø; D(2,2);

440    PRINT AT 3,7; C$(2,1 TO 12)

450    PRINT AT 2,20 C$(4,1 TO 12)

460    PRINT  TAB 20;C$(5,1 TO 12)
```

470   FOR N = 1 TO 10

480   LET X(1,N) = 30.* (A(1,N)/XMAX)

490   LET X(2,N) = 30.* (A(2,N)/XMAX)

500   LET Y(1,N) = 15.* (B(1,N)/YMAX)

510   LET Y(2,N) = 15.* (B(2,N)/YMAX)

520   NEXT N

530   FOR N = 1 TO 10

540   PRINT AT (19-Y(1,N)), X(1,N); "*

550   PRINT AT (19-Y(2,N)), X(2,N); "0"

560   NEXT N

570   PRINT AT 1,12; C$(3,1 TO 12)

INPUT

Plot Name: PLOT TWO

XMAX: 10.

YMAX: 500.

| X NAME: | TIME, MIN | | * DATA NAME: | OLD * |
| Y NAME: | TEMP R | | 0 DATA NAME: | NEW 0 |

* DATA:	1	ENTER	50	ENTER				
	2		100					
	3		150					
	4		160					
	5		170					
	6		200					
	7		250					
	8		300					
	9		400	0 DATA	5	ENTER	100	ENTER
	10		500		6		120	
0 DATA:	1		20		7		130	
	2		30		8		140	
	3		40		9		200	
	4		50		10		300	

107

BASIC PROGRAM NUMBER 14: PLOTTING FOR TWO DATA SETS

PLOT TWO

NEW *
OLD ∅

TEMP R

500 —
400 —
300 —
200 —
100

TIME, MIN

2   4   6   8   10

```
1Ø    REM "PLOT KP"

2Ø    DIM A(2,1Ø)

3Ø    DIM B(2,1Ø)

4Ø    DIM C$(5,12)

5Ø    DIM D(2,1Ø)

6Ø    DIM X(2,1Ø)

7Ø    DIM Y(2,1Ø)

8Ø    PRINT"INPUT PLOT NAME"

9Ø    INPUT C$(3,1 TO 12)

1ØØ      PRINT"INPUT X NAME"

11Ø      INPUT C$(1, 1 TO 12)

12Ø      PRINT"INPUT Y NAME"

13Ø      INPUT C$(2, 1 TO 12)

14Ø      PRINT "INPUT NUMBER OF DATA POINTS"

15Ø      INPUT Y

16Ø      PRINT "INPUT TEMP, R; INPUT IN KP FOR * DATA"

17Ø      FOR N = 1 TO Y

18Ø      INPUT A(1,N)

19Ø      INPUT B(1,N)

2ØØ      PRINT A(1,N); B(1,N);

21Ø      NEXT N

22Ø      PRINT "INPUT TEMP, R; INPUT KP FOR Ø DATA"

23Ø      FOR N = 1 TO Y

24Ø      INPUT A(2,N)
```

```
250     INPUT B(2,N)

260     PRINT A(2,N); B(2,N)

270     NEXT N

280     PRINT"INPUT * DATA  NAME"

290     INPUT C$(4,1 TO 12)

300     PRINT "INPUT Ø DATA NAME"

310     INPUT C$(5,1 TO 12)

320     LET D(1,1) = Ø

330     LET D(1,2) = 1/1000

340     LET D(1,3) = 1/500

350     LET K = -25

360     FOR N = 1 TO 9

370     LET D(2,N) = K

380     LET K = K + 5

390     NEXT N

400     CLS

410     LET DISTA = (D(1,2)/D(1,3))* 25 + 5

420     LET DISTB = 30

430     PRINT AT 19,5; "I"; TAB DISTA; "I"; TAB DISTB; "I";

440     PRINT AT 20,5; "Ø"; TAB DISTA; D(1,2); TAB 28; D(1,3);

450     PRINT AT 21, 18; C$(1,1 TO 12)

460     LET K = 2

470     FOR N = 2 TO 9

480     LET K = K + 2

490     PRINT AT K,4; "-";
```

```
500    NEXT N

510    LET K = 4

520    FOR N = 1 TO 9

530    PRINT AT K,0; D(2,N)

540    LET K = K + 2

550    NEXT N

560    PRINT AT 3,7; C$(2,1 TO 12)

570    PRINT AT 2,20; C$(4,1 TO 12)

580    PRINT TAB 20; C$(5,1 TO 12)

590    FOR N = 1 TO Y

600    LET X(1,N) = 5 + 25 * (1/A(1,N))/D(1,3)

610    LET X(2,N) = 5 + 25 * (1/A(2,N))/D(1,3)

620    LET Y(1,N) = 14 + (B(1,N)/2.5)

630    LET Y(2,N) = 14 + (B(2,N)/2.5)

640    NEXT N

650    FOR N = 1 TO Y

660    PRINT AT Y(1,N), X(1,N); "*";

670    PRINT AT Y(2,N), X(2,N); "0";

680    NEXT N

690    PRINT AT 1,12; C$(1, 1 TO 12)
```

INPUT

PLOT NAME: AMMONIA FORMATION

X NAME: 1/T, R

Y NAME: IN KP

NUMBER POINTS: 4

      * DATA NAME: * NH3
      ∅ DATA NAME: ∅ 2 NH3

* DATA: 537, 6.78
        1000, -2.2
        2000, -8.1
        5000, -11.78

∅ DATA: 537, 13.55
        1000, -4.4
        2000, -16.19
        5000, -23.55

```
                    AMMONIA FORMATION
                              *  NH3
            IN KP              ∅  2 NH3
     -25  -
                    ∅
     -2∅  -

     -15  -
                         ∅

     -1∅  -    *
                    *

      -5  -
                         ∅
                         *
       ∅  -

       5  -
                                        *

      1∅  -
                                          ∅

      15   I                I                I
           ∅              .001             .002
                              1/T, R
                               112
```